Baumeister, Michael

Poultry as a hobby

POULTRY
AS A HOBBY

The Royal Palm is an especially beautiful turkey variety.

Michael Baumeister & Heinz Meyer

POULTRY
AS A HOBBY

Translated by
WILLIAM CHARLTON

Poultry as a Hobby is a translation of *Geflügelhaltung als Hobby,* © 1985 by Falken-Verlag GmbH, 6272 Niedernhausen/Ts., FRG. English translation © 1988 by T.F.H. Publications, Inc. New material has been added to this English-language edition, including but not limited to additional photographs. Copyright is also claimed for this new material.

Drawings by GABRIELE HAMPEL, Kelkheim/Ts.
Photographs by DR. HERBERT R. AXELROD; MICHAEL BAUMEISTER; GARY HERSCH; P. LEYSEN; HEINZ MEYER; HELMUT NEUBÜSER; HANS REINHARD.

The publishers wish to express appreciation to HORST SCHMUDDE *for his valuable guidance in the preparation of this book.*

Distributed in the UNITED STATES by T.F.H. Publications, Inc., One T.F.H. Plaza, Neptune City, NJ 07753; in CANADA to the Pet Trade by H & L Pet Supplies Inc., 27 Kingston Crescent, Kitchener, Ontario N2B 2T6; Rolf C. Hagen Ltd., 3225 Sartelon Street, Montreal 382 Quebec; in CANADA to the Book Trade by Macmillan of Canada (A Division of Canada Publishing Corporation), 164 Commander Boulevard, Agincourt, Ontario M1S 3C7; in ENGLAND by T.F.H. Publications Limited, Cliveden House/Priors Way/Bray, Maidenhead, Berkshire SL6 2HP, England; in AUSTRALIA AND THE SOUTH PACIFIC by T.F.H. (Australia) Pty. Ltd., Box 149, Brookvale 2100 N.S.W., Australia; in NEW ZEALAND by Ross Haines & Son, Ltd., 18 Monmouth Street, Grey Lynn, Auckland 2, New Zealand; in SINGAPORE AND MALAYSIA by MPH Distributors (S) Pte., Ltd., 601 Sims Drive, #03/07/21, Singapore 1438; in the PHILIPPINES by Bio-Research, 5 Lippay Street, San Lorenzo Village, Makati Rizal; in SOUTH AFRICA by Multipet Pty. Ltd., 30 Turners Avenue, Durban 4001. Published by T.F.H. Publications, Inc. Manufactured in the United States of America by T.F.H. Publications, Inc.

CONTENTS

PREFACE

In a speech to the American president in 1855, the Indian chief Seattle said: "What would man be without animals? If all animals disappeared, then man would die of a great spiritual loneliness. Whatever happens to animals soon happens to man as well. All things are interconnected. Whatever befalls the earth also befalls the sons and daughters of the earth."

Since these words were spoken, many species of animals have left our globe forever. Conservation organizations try to stop the plunder of nature. Political parties see new value in the concepts of nature and ecology, and turn themselves — depending on their views — more or less intensively to these "new" areas. In the general population as well, the desire for a way of life aware of and close to the environment becomes stronger and stronger.

Before all of these movements arose, poultry-breeding hobbyists attempted to introduce nature in the form of domestic-animal breeding into the cold, concrete deserts of the cities. With breeding facilities and poultry shows they continue to present an ancient, living cultural heritage of mankind: domestic poultry, tamed and formed by the hand of man. For many children, the activities of poultry clubs made possible the first direct man-animal contact in their lives.

At the present time, in which one increasingly returns to natural and ancient values, the idea of keeping and breeding their own chickens, ducks, geese, turkeys, or guinea fowl is growing within many people. The fresh egg or the pleasure of daily contact with feathered companions is, for many, sufficient reason to finally realize the dream of keeping their own poultry.

Let us hope that the return to the living creature wins out over the cliche of the domestic rooster's crowing as a "noisy disturbance," and that man recognizes that these natural sounds of a fellow creature enrich his own inner emotional life! Man should never forget that he comes from nature and is himself a part of nature!

Through the hobby of poultry breeding he can bring a part of the natural world into, or at least close to, his home, and can realize to a certain extent the deeply embodied emotional feeling for nature. Perhaps he will also be granted a little happiness in this way.

IDEA AND MOTIVATION

So, dear reader, you have taken the first step on the way to becoming an active poultryman. You have certainly already toyed for a fairly long time with the idea of enriching your leisure hours with the addition of a few beautiful birds, which will perhaps even yield a small profit for the household. Now you have purchased this specialized book in order to establish the theoretical foundation for your plan. Many motives could have entered into your decision. Is it the intention of sensibly using an existing shed on your land, or the desire to populate an available orchard with beautifully colored poultry? Or perhaps, to provide your household, and perhaps a few other people, with the tasty products of domestic poultry, without the "producers" having to put up with a degrading existence without daylight in cramped cages? With a small flock of poultry in your own garden, providing for the birds' natural requirements, you will make a worthwhile contribution to the animal world.

Perhaps you also have fond memories of your childhood, which involved small animals kept by your grandparents or with the animals on a relative's farm, where you were allowed to spend several weeks' vacation. In keeping poultry you now have the opportunity to realize your childhood dreams and to foster your contact with all living creatures. But poultry keeping is not as easy as it looks: each living creature has specific requirements which must be taken into consideration under all circumstances.

In this book the demands of the different kinds of poultry with respect to housing, run, climate, feed, and care will be discussed in detail. The results you expect from keeping poultry balance the expense and the amount of time needed for the birds' care. In setting up your own poultry facility, to keep the costs as low as possible you should consider the proven guidelines recommended for the various kinds of poultry in this book. Given a little mechanical skill, you can build most of the equipment yourself and suit it to existing conditions.

If you have not yet decided on a particular poultry breed, you should consider what you expect from your future leisure activity. For instance, if you value only the maximum possible egg production, then you should choose a hybrid breed. Here too, there is now a degree of choice in plumage and eggshell color. If the available space is very limited, then a productive bantam breed is recommended. If you mainly wish to supply the kitchen with tasty poultry flesh, then one of the fast-growing meat breeds are recommended, unless you wish to raise seasonal fattening hybrids. If a small body of water with an adjoining meadow area is available, then the prospects of keeping waterfowl are very good. Turkeys also require — besides special feed for rearing — large amounts of greenfood. Guinea fowl are eager foragers if sufficient run space is available.

One aspect, which has up to now been considered by only a few poultry fanciers, is the cultural significance of various indigenous poultry breeds. Many are already extinct as a result of displacement by "more modern" breeds, or merely languish in small numbers, neglected by breeders. Too little consideration has been given to the fact that, with the disappearance of the breeds adapted to regional conditions, the specialized, once economically significant genes are also lost. The preservation of many poultry breeds is due only to the efforts of purebred-poultry breeders, who, as a result of their fancy, have preserved the multicolored diversity. For this reason, if you do not strive for maximum profitability in your poultry keeping, you should attempt, by purchasing poultry breeds indigenous to your region, to preserve them.

For the fanciers of the unusual, domestic poultry also offers diversity, ranging from the large, bulky Brahma chicken or the small, short-legged Chabo to the elegant long-tailed breeds Yokohama, Sumatra, and Phoenix. An unusual form of the latter — called Onagadori in its Japanese homeland — can produce ornamental tail feathers up to 16 meters long under specialized housing conditions! In striking contrast to this abundance of feathers stand the Rumpless Fowl and the Araucana — the latter is the only chicken breed that lays eggs with turquoise-blue shells. Silkie Bantams have lost the rigid structure of their feathers; their plumage is hairy and soft as silk.

Waterfowl also include breeds of unusual appearance, such as the Sebastopol goose, whose wing-covert feathers are elegantly twisted like iridescent curls and droop from the shoulders and back almost to the ground. With ducks, there are breeds with a feather crest on the head; the numerous chicken breeds with feather crests and beards, or both, have almost gone unmentioned. In addition, there are also numerous color and marking varieties in all poultry species and breeds.

DOMESTICATION AND THE ANCESTRY OF POULTRY

The Unresolved Questions of Domestication

Originally, domestic-animal researchers adopted the view that particular domestic animals, such as dogs, goats, rabbits, geese, ducks, chickens, and others, in each case derived from several wild species. Modern domestic-animal zoology, which is based on the idea of the biological species, maintains that all the various forms of a domestic animal originated from a single wild species, and that they are not species in their own right. In attempting to determine which species contribute to the ancestors of our domestic animals, external appearance (so significant in distinguishing wild species) can be relied upon only to a very limited extent; indeed, extreme dissimilarity in appearance and diversity in shape are what distinguish the various domesticated forms of our animals. Consider, for example, the approximately 150 chicken breeds in all their diversity! The long-tailed Phoenix, Yokohama, and Sumatra and the tailless Araucana and Rumpless Fowl come to mind. Additional contrasts are formed between the giant Brahma chicken and the tiny bantam, as well as by the ornamental plumage of the Italian rooster compared to the hen-feathered Sebright rooster. All of these contrasting items demonstrate the diversity within one group of domestic animals.

The Concept of an Animal Species

If one proceeds from the definition of the species proposed by the French naturalist Cuvier in 1829, who viewed an animal species as a natural reproductive community with unrestricted mate selection, then errors that occur because of similarities in form between wild and domestic animals fall by the wayside. Through the mixing of genes during reproduction, the species population attains a relative genetic purity, whereby one can recognize the species as something common to all individuals that are descended from one another, produce fertile offspring, and share fundamental traits. Despite genetic purity, one must also consider the space-time aspect, which is capable of changing the species' traits without all individuals of the species in question being affected by it. As a result, the characteristics of a species vary without, however, interrupting its ability to reproduce. This situation is well known with the Mallard. Pure Mallards hardly exist anymore; as a rule, they are hybridized with feral or escaped domestic Mallards. For this reason, variations in form and coloration result, even though reproductive continuity persists.

The Domesticated Animal in Zoological Systematics

Since it was recognized early on that a species displays a marked intraspecific variability (differentiation), as early as 1861 a Briton named Bates proposed the concept of the "subspecies." According to this idea, a subspecies is a geographically isolated group of local populations that differs genetically and systematically from other subspecies. All subspecies taken together make up the species. Systematically, a wild animal species is designated by the so-called binomial (two name) nomenclature; that is, the animal has a genus and a species name. On principle, the science of biology classifies animals according to the hierarchy of phylum, class, order, family, genus, and species in the overall zoological system. If the subspecies name is then added to the binomial nomenclature, one speaks of a trinomial (three name) nomenclature. An example is *Gallus gallus bankiva*. *Gallus* = generic name, *gallus* = specific name, *bankiva* = subspecific name. The three names distinguish the Javanese subspecies of the Red Junglefowl. The species Red Junglefowl breaks down into a total of five subspecies: the Cochinchina Red Junglefowl (*Gallus gallus gallus*), the Indonesian Red Junglefowl (*Gallus gallus murghi*), the Tonkinese Red Junglefowl (*Gallus gallus jabouillei*), the Burmese Red Junglefowl (*Gallus gallus spadiceus*), and the Javanese Red Junglefowl (*Gallus gallus bankiva*). In their entirety the five subspecies form the species commonly known as the Red Junglefowl. Accordingly, the species is conceived as all the subspecies collectively. In captivity, all of these subspecies will form a natural reproductive community, given unrestricted mate selection, even though they differ in particular characters. In the wild, interbreeding of subspecies occurs only rarely, since they are separated from each other by geographical barriers. Another wild chicken, the Grey Junglefowl, can be distinguished very easily from the Red Junglefowl. It has the scientific name *Gallus sonnerati*. From this one can recognize that it belongs to the same genus as the Red Junglefowl but is a different species. If it belonged to the Red Junglefowl species, then it would have the name *Gallus gallus sonnerati.* Since there are no subspecies of the Grey Junglefowl, the binomial suffices.

Domestic animals do not themselves represent separate species, but instead are always classified with the appropriate ancestral species with which they would form a reproductive community producing fertile offspring if given the opportunity for free mate selection. In other words: the domestic chicken can mate with a Red Junglefowl and produce viable offspring.

Scientific names are reserved for fowl occurring in the wild; man-made varieties are never "species." Thus domestic animals do not have their own scientific names; instead, they are classified as "forma" (= forms of the ancestral species). If an exact domestic-animal name is lacking, then one instead uses the designation *domestica.* Accordingly, the domestic chicken carries the designation *Gallus gallus* forma *domestica.* The name *Gallus gallus* indicates descent from the Red Junglefowl.

Since, however, considerable variation in regard to appearance and production also occurs within a domestic animal form (forma *domestica*), for further distinctions one introduces the concept of the "breed." In a sense, the breed of a domestic animal is analogous to the subspecies of a wild animal.

Within the breeds of domestic animals we differentiate according to type (large, medium, and dwarf forms) and color variety.

We recognize the following main categories of color varieties: wild color, silver-wild color, penciled, laced, spangled, lacquered, Columbian, solid, pied, barred, and tricolored with its variations.

The Reasons for Domestication

The reasons that prompted mankind to domesticate animals can be theoretically inferred from a combination of natural-historical findings and cultural-historical research. In this context one must differentiate between *taming* and *domestication.* By *taming* is meant only the keeping of wild animals in captivity as pets (for example, ornamental poultry today), while *domestication* is contingent upon the genetic change of the wild animals. The true domestic animal was developed as a result of this process. Without doubt, however, both processes are closely tied to each other, whereby taming always precedes domestication. In the initial phase, the domesticator had no idea of the value his animal products would have for future generations. The tamed wild fowl laid but few eggs, and the wild cattle gave hardly any milk. The animals he kept had significance only as a living meat reserve. Thus the principal reason for taming can be considered to be a delight in animals. This process is comparable to keeping tropical fish, a parrot, or a hamster nowadays: the animals yield no profit but have a very positive effect on man's psychic domain. A bit of nature and a bit of life in the home: this must often have led to the domestication of animals. As a result, nomadic wandering was no longer possible. Domestication goes hand in hand with the establishment of permanent settlements. In places where the transition from hunting and a nomadic existence to agriculture was accomplished, the process of domestication also took place. As a result of animal husbandry, man had to protect and feed his domestic animals.

Changes in Animals through Domestication

Wild animals that were domesticated underwent substantial alterations in the process. These changes include the so-called problem of allometry and changes in skeleton, body covering, circulation, musculature, reproduction, hormonal balance, life cycle, and behavior. The allometric problem refers to changes in body size and proportions. This is evident in particular in chicken breeds. We recognize heavy, medium-weight, light, and bantam breeds, as well as the true bantam breeds. Loss or increase of pigment is exhibited in the feathers; we then speak of so-called whites or blacks, respectively. In Silkie Bantams, shredded feathers are added, as well as a new skeletal form with five toes. In contrast to the wild form, the musculature displays heavy fat deposits and less endurance. Because of the greater productivity of domestic animals, their metabolism is changed, as are the metabolic organs. The fertility of domestic animals is increased greatly in comparison to that of any wild animal. As a rule, domestic animals become sexually mature sooner, and sexual maturity is often reached even before growth is completed. Our young roosters, which tread (that is, mate) before they are fully grown, are a good example of this. The reproductive cycle changes in the same way. Domestic animals are generally able to reproduce independently of the season. Wild chickens produce their broods in the spring. Afterwards the hen lays no more eggs. The rooster's reproductive activity also dies down. Externally this is recognizable by the nonbreeding plumage. The peafowl can also serve as an example of this. In contrast, our domestic hens lay eggs throughout the entire year. The rooster no longer has a nonbreeding plumage, being sexually active throughout the entire year.

In the nervous system, the brain capacity is of particular interest: it decreases in the domesticated forms. In behavior, it can lead to a decline in instinctive actions; on the other hand, hypertrophy (exaggerations) can occur. For example, fleeing distances have decreased and aggressive behavior has been reduced. In addition, behavioral habits can differ; for example, courtship displays no longer function, so that mating often takes place without courtship. Not infrequently, the bodily structure is altered so greatly that a behavior sequence can no longer be carried out properly: Araucanas and Rumpless Fowl have virtually no tail, which normally functions to stabilize the body during mating.

The Ancestors of Domestic Chickens

In excavations of 4,000-year-old ruins in India, archaeologists found small clay figures of massive birds that looked like chickens. In Mohenjo-Dara—an ancient town of the Indus high culture—such a bird perched in front a food dish was unearthed. If one compares figures of this kind with recent portrayals of chickens by primitive peoples, then an enormous parallelism can be established. Since massive and heavy chicken bodies are shown here,

One of the crested chicken breeds, a Polish hen with her chicks.

while the Red Junglefowl indigenous there has a light body structure, it can be assumed that the clay figures represented chickens that were already bred and domesticated for a long time. This theory is further corroborated in that one knows from domestic-animal research that the size of domestic animals decreases in the beginning, and only later—during the continued domestication process—can an enlargement of body size be attained.

Chinese writings mention the domestic chicken as early as 2800 B.C. It can be assumed that it had already been tamed and domesticated earlier in various locations. It is also debated whether other wild fowl besides the Red Junglefowl were among the ancestors of the domestic chicken.

The Ancestors of Domestic Geese

The progenitor of our domestic geese is the Greylag Goose (*Anser anser*). Its homeland is in Europe and Asia. The beginnings of its domestication can no longer be determined. It is said, however, to have been first kept about 3500 years ago in Mesopotamia. Other sources shift its initial domestication to Europe and China. In the history of civilization, geese drew attention to themselves in ancient Rome, where they announced the approach of enemies with their cackling. In terms of the numbers kept, geese are, so to say, step-children in comparison to chickens. In breeding, one principally values weight, readiness to breed, and egg production.

Chinese geese occupy a special position. They did not arise from the Greylag Goose, but instead from the Swan Goose (*Anser cygnoides*) of Asia. The Steinbacher Game goose and the Celler goose have the blood of both Greylags and Swan Geese in them. Authenticated facts clearly show that not all domestic-animal forms always originated from a single wild species, as was claimed for a long time.

The Ancestors of Domestic Ducks

The duck was domesticated later than the goose. Apparently the need for a place to swim was the reason for the late domestication. The ancestor of breed ducks is the Mallard (*Anas platyrynchos*). From time to time the participation of other duck species in the domestication process is also debated. In breeding, one selects, above all, for fattening ability (for example, the Pekin duck), for egg production (the Campbell duck), and for beauty.

The familiar Muscovy originated from the wild form of South America's Muscovy Duck (*Cairina moschata*). Thus our domestic ducks stem from at least two wild species.

The Ancestors of Domestic Turkeys

The Common Turkey comes from southern North America and Central America. When the white man conquered the new continent, he met with whole flocks of domesticated turkeys; thus the domestication of the turkey was begun before the time of Columbus. On the basis of archaeological finds, one can assume that domestication began about 500 B.C. The ancestral form of the turkey is the species *Meleagris gallopavo* with its seven subspecies. Moreover, the Mexican Turkey (*Meleagris gallopavo gallopavo*) is probably the principal ancestor. From the remaining six subspecies (Eastern Turkey, Rio Grande Turkey, Merriam's Turkey, Florida Turkey, Sierra Madre Turkey, and Gould's Turkey), primarily the Eastern Turkey took part in turkey breeding. Around 1570, the turkey was systematically bred in Germany, and soon it replaced the Common Peafowl (*Pavo christatus*), which hitherto was kept for epicurean roasting. Similarly, holiday feasts in America feature turkeys.

The Ancestors of Domestic Guinea Fowl

In ancient Greece—therefore more than 2000 years ago—the guinea fowl was kept as domestic poultry. Later it reached Italy, principally because of its tasty flesh. With the fall of the Roman Empire and highly developed Roman agriculture, the guinea fowl also disappeared as domestic poultry. The Portuguese again brought it from Africa to Europe approximately 1000 years later, where it became more and more widespread after the eighteenth century.

Of the various genera of guinea fowl, only the Helmet Guineafowl is the progenitor of our domestic guinea fowl. In antiquity, subspecies other than the present-day *Numida meleagris galeata*, which comes from the West African savanna, composed domestic guinea-fowl flocks. It is bred in diverse color varieties, primarily because of the tasty flesh and the delicious eggs.

REQUIREMENTS FOR KEEPING POULTRY

Besides the space required for the poultry, the effects on the immediate vicinity must be considered to a greater extent today than in the past. Unfortunately, court verdicts time and again have favored neighbors who felt that keeping birds produced a health hazard. For this reason, it is advisable in all cases to obtain the neighbors' consent before starting to keep poultry, and to check with local zoning authorities as well.

Necessary Space

We must tailor the kinds of poultry to the available plot of land. The planned run should be protected from damaging surface winds or sheltered with appropriate plantings. If the area is small, rush mats or something similar can be used. A run sited open to the southeast is ideal. Because of the noise the poultry make, the coop and run should, of course, be located far away from neighbors' houses. On the other hand, the installation should be as easily accessible as possible from one's own residence.

Chickens and Bantams

The amount of space needed for chickens of normal size — approximate figures for planning — can be taken from the accompanying table.

Smaller coops are preferably built of wood, and, using uprights, are raised about 50 centimeters above the ground. This space — open only on the south side — is available as an extra scratching area.

With bantams, which in several breeds can also provide very high production, approximately one third more birds can be kept for each square meter of coop space. With particularly small breeds such as the Holland Bantam, Old English Game Bantam, or Modern English Game Bantam, as well as with the short-legged Chabo, the run need be only about one third of the size given previously. With commercial bantam breeds, however, a somewhat larger run should be provided.

If a grass run cannot be provided because of limited space, we can provide a good alternative with alternating runs or a small litter-covered, possibly roofed run with an adjoining grassy area that is only used periodically. The alternating runs are best set up in such a way that by opening the appropriate run door from the coop they can be reached as desired. After a maximum period of use of eight days, the vegetation must be given at least the same amount of time to recover, aided by cultivation measures (cutting the worn grass, fertilization, watering, etc.). Do not use insecticides!

Small runs must have good water drainage; if necessary, the ground should be dug out to a depth of about 30 centimeters and a drainage layer at least five centimeters thick — consisting of coarse gravel or cinders — should be installed. If the soil is clay or loam, then only sand should be used for the drainage layer. Contaminants such as droppings and spoiled greenfood should be removed regularly, and the soil should be turned over monthly after spreading on quicklime. At the start of the new season, in early spring, the top layer of soil should be replaced. If the necessary cultivation measures are observed, chickens in these small runs are just as healthy as on grassy areas.

Another possibility is a small covered run, at least twice the size of the ground space of the coop and which is provided with a layer of litter at least 30 centimeters deep. This litter should preferably consist of straw; between the ground (or the solid surface) and the litter one places an insulating layer — consisting of a five to ten centimeter layer of wood shavings, or, better yet, of shredded tree bark. The latter is being increasingly used as a mulch in public plantings and parks, and also has a positive effect on decomposing microorganisms in the permanent or deep litter of small, littered runs. If the deep litter has been properly applied and the microorganisms have become established, then droppings will be rendered harmless by the microorganisms. The maintenance of these runs will then merely consist of the constant replenishment of the litter, which, incidently, can also consist of lightly composted woodland leaf-litter.

It is always a pleasure to watch the chickens busily working over this deep litter. If the litter should happen eventually to become too compacted, then we must loosen the bedding with a pitch fork and work in some straw. Depending on

Housing Requirements Depending on Number of Birds

Number of Birds	*Coop Area (m^2)*	*Window Area (m^2)*	*Run Area (m^2)*
3	1	0.2	50
5–7	2	0.4	100
8–10	3	0.6	150
12–15	5	1.0	250
20–25	8	1.5	400

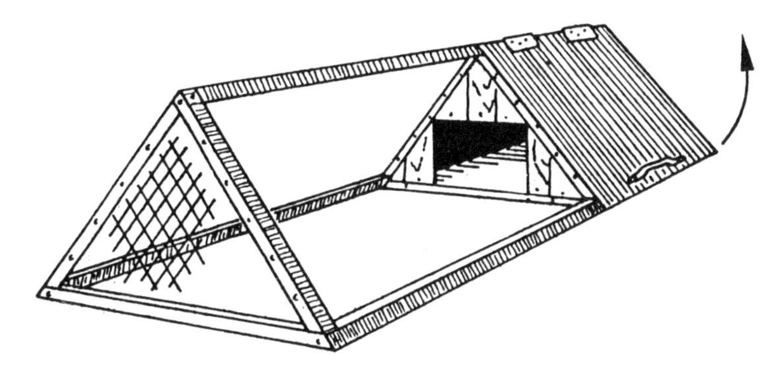

Portable enclosure with a small shelter for a brood hen with chicks or for several youngsters. Also suitable for artificially rearing a gosling until it is feathered. Protects against losses from predators.

need, the litter is changed once or twice a year; two to three buckets of the old litter should be saved for "inoculating" the new. Disinfection of the cleaned run with one of the agents (preferably with one of the so-called biological preparations) available commercially is advisable.

In these runs, greenfood — according to the season — should be offered to the poultry, either shredded in hayracks or hung up in bundles. If conditions permit, it is beneficial to have a fenced-in grassy area adjoining the small run. Here the birds can supply themselves with the greenfood and the supplementary food it contains, such as worms and insects. So that the vegetation is not destroyed, the fowl should only have access for short periods of time. When creating combined facilities of this kind, one is not bound by any fixed rules; here no limit is placed on the poultry keeper's "ingenuity." There are runs, for example, that can be reached by the fowl only through a long tunnel constructed of wire mesh.

Geese and Ducks

Considerably different requirements for the run are needed for keeping ducks and geese. Ducks do not require as much pasture as geese, but both need water for bathing. If only a few birds will be raised for personal requirements, then a water basin about 50 centimeters in diameter and 10 to 15 centimeters deep placed in the run is sufficient for at least partially feathered birds. Small ducklings or goslings — which is what the chicks of a goose are called — when raised naturally are provided with the necessary protection for their down plumage from the oiled plumage of the mother, and thus can already bathe and swim safely. With artificial rearing the plumage oil is not present at first; therefore, until the lower half of the body is feathered, the youngsters should be given water only in drinking troughs. The rims of the drinking troughs should, however, be deep enough that the bill can be completely submerged. If they can later be offered a small bathing area in a grass-covered run, this will serve to promote their well-being. This is true for most duck breeds, but Muscovies occupy an exceptional position. Like geese, they are not as dependent on a bathing opportunity.

Geese should be provided with a run area of 150 square meters of fresh pasture per bird. Small goslings or ducklings that are being artificially reared can at first be kept in a portable pen, which is also enclosed with wire on top to protect against cats and birds of prey. Moisture from above should be stopped by a suitable covering until the ducklings and goslings are fully feathered, side winds by placement of a screen, and ground cold with a well-littered, easily accessible hutch. If only a very limited area is available for keeping geese, then it is advisable to at least divide it into quarters so that the grass can recover after grazing. If only a few birds are present, a portable run — which can be moved as soon as the grass has been eaten — is also recommended. The fencing for birds no longer threatened by predators need only be one meter high with geese and ducks (except for the duck breeds that can fly). For partitioning, even a height of 50 centimeters is adequate, particularly with ducks. For ducks, a pasture area of 15 square meters must be allowed for each bird; Muscovies, which eat more grass, should each have twice this area.

If we cannot offer the ducks a grass run, then an enclosure with at least one square meter per duck in addition to the coop area of one square meter per four birds is sufficient. In this case, the run should also be provided with straw. Additionally, for ducks, as large a distance as possible between the food and water containers should be provided, since otherwise the food will be carried in the bill to the water and will spoil. If ducks are not being kept for commercial purposes, it is advisable to feed a supplement of greenfood in the enclosures as well; besides saving feed, this also maintains the health and appetite.

Turkeys and Guinea Fowl

Turkeys and guinea fowl will naturally roam very far if given an unlimited run and find much of their own food. Today, however, this can be provided for the birds only in the rarest cases; usually only a limited run is available. In any case, a grass run is recommended for turkeys. Guinea fowl, with appropriate feed-

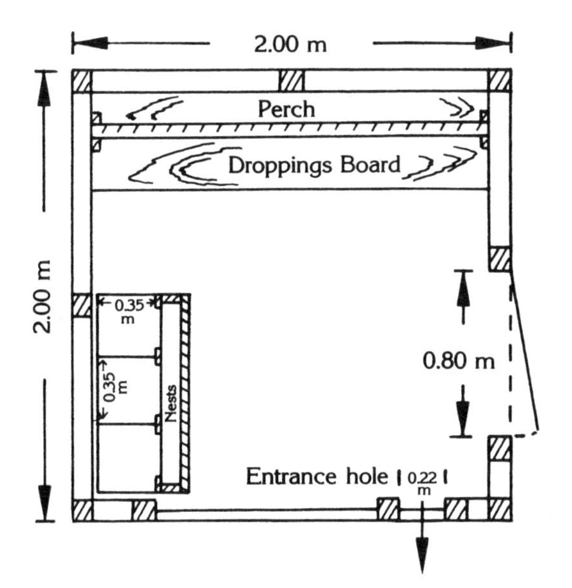

Floor plan of a chicken coop spacious enough for 12 to 15 birds. Requires a foundation that must be insulated against moisture rising from the ground.

ing, can also be kept in enclosures without vegetation.

Adult turkeys have no unusual requirements with respect to accommodations. A simple wooden coop is quite sufficient; drafts must, however, be prevented. Turkeys cannot tolerate wetness in the run; perfect drainage of the run area is most important. One square meter of coop area can suffice for one turkey; the fence should reach a height of two meters. With breeding stock, 100 square meters should be allowed for each bird. The delicate youngsters must always be given runs that have not been used by poultry for at least a year. In this way, the danger of an infection from the agents of blackhead disease can be kept to a minimum. Turkey youngsters, on account of the high risks, should not be acclimated to the outdoor run until they have reached an age of about eight weeks. With intensive rearing, the hygienic requirements as well as the necessary warmth are easier to guarantee.

Guinea fowl animate any poultry facility because of their temperament, but the amount of noise they produce could cause problems with neighbors. Because of their characteristic nervousness, when keeping them in enclosures it is advisable to cover the run with netting or light chicken wire. The same coop and run dimensions given for chickens should be used. Keeping guinea fowl with other kinds of poultry is only possible in very large runs, since guinea fowl are very quarrelsome with other poultry. With limited runs, keeping guinea fowl separately is advisable.

The Coop and its Furnishings

Chickens and Bantams

If we expect the greatest possible production from our chickens, then, in addition to an optimal diet, we must offer them the most suitable accommodations possible. That is, the birds must in rainy, cold, and stormy weather have shelter that meets their minimum requirements. In earlier times, the opinion was held that the birds required accommodations that were as warm as possible, so that laying was not interrupted. For this reason, they were often kept in a stall in warm cattle barns. When kept in this way in damp air, which additionally was usually deficient in oxygen, pathogens could spread and wipe out entire flocks. With commercial importance, poultry keeping was given proper attention, so that coops suitable for chickens were developed.

The following attributes of the coop are critical for the chickens' well-being: light, dry, well-ventilated and at the same time free of drafts. The simplest possible furnishings, which facilitate cleaning and offer vermin little opportunity for entry, should be provided for the interior of the coop. Suggestions for furnishing and constructing the coop can be drawn from the various illustrations in this book. Very well-constructed coops made of prefabricated wood suitable for small poultry may be found on the market today. These coops are insulated with modern building materials, lined on the inside with particle board, and also have a clean and neat appearance. Unfortunately, the price is correspondingly high. Whoever has time and a little mechanical skill can also try to build the necessary coops, and in addition give the entire installation an individual character.

A concrete foundation and a well-insulated floor should always be provided for the sake of maintenance and ease of cleaning. Then too, no rats or other vermin can enter the coop through the floor. The walls should preferably consist of a wooden framework, which is covered on the inside with particle board and on the outside with tongue-and-groove boards. Fiberglass mats are better than styrofoam sheets as insulating material, since mice are particularly fond of gnawing apart the latter. (These little gnawers can squeeze through even the smallest gaps to reach the hollow space between the walls.) The roof must be covered with a double layer. The space between the coop's ceiling and the roof (with a lean-to roof) should be about ten centimeters; equipped with vent openings, it is often used for the ventilation.

Elevation of the coop already shown. A grating installed above the window ensures sufficient ventilation.

It would be ideal if the coop temperature for chickens did not reach much above 20° C. in summer or below 5° C. in winter. For egg production, a sufficient "light period" is critical. In the time of year with low light levels, the light source in the poultry coop should be turned on so that approximately 14 hours of light are available to the chickens. The light source should be placed over the food containers in the flocking area, so that the birds are kept active. Roosts and nests can safely lie in semi-darkness; they will be visited as needed under these conditions as well. If the lighting is suitable and the water is prevented from freezing, good egg production can be expected even in deepest winter. In severe cold, the eggs should be removed from the nests several times a day, so that the quality remains satisfactory.

The air quality in the coop is of great importance for the well-being of the poultry. The amount of air required and the corresponding "air space" per chicken must be considered in planning the coop. Depending on the size of the flock, a coefficient of 0.5 to 0.7 cubic meters of air space per chicken can be allowed. Nevertheless, properly functioning ventilation of the coop is mandatory. In this respect, the following arrangements work well:

- Movable shutters or slides are suitable for the space between the ceiling and the roof (with lean-to roofs), with which the air supply can be regulated according to need.
- Smaller coops with gable roofs should also be equipped with slides

or shutters on the side. In this case, however, the ventilation openings are located in the gable wall.

- The windows should be constructed and installed in such a way that they can be completely or partly replaced with screens in the warm season.

Although, in most cases, permanent coops built of wood on solid foundations (these must extend below the frost zone — usually about 80 centimeters) are used for noncommercial poultry keeping, solid construction or installation in an existing building should not be overlooked during planning. Masonry coops must be particularly well-protected against dampness and must be provided with perfectly functioning ventilation. In poorly ventilated and slightly damp coops, respiratory illnesses are inevitable.

With new construction or the remodeling of existing structures, sufficient window area and ample air circulation must be provided; the cracks and nooks (hiding places for parasites and vermin) which are usually present must be completely and permanently sealed. It is best to face wooden structures with particle-board or hardboard sheets. The latter will fit snugly if one wets the fitted sheets on the rough side that becomes the inside of the wall and nails them to the wooden frame in this condition. Existing floors should also be prepared in the same way, and, if necessary, sealed against dampness.

If the coop space appropriate to the expected number of chickens is available, then we should now assemble the interior furnishings. The basic furnishings consist of a dropping board and perches, nests, and food and water containers. If a dust bath is available in summer in the run for hygiene, this does not have to be allowed for in the coop space. It is then sufficient, during the cold time of year, to place a 20 centimeter high, 60 centimeter wide, and 80 centimeter long wooden box with sand and healthful supplements in the flocking area. This should be set up in a sunny spot and should be enriched with a small amount of wood ash, as well as with tobacco dust and flowers of sulfur, if available, as a preventive measure against parasites, or otherwise with some insecticide powder (for use on birds!).

Numerous reasons for the installation of a dropping board under the perches can be put forward. In the first place, flocking space will be gained, the litter stays dry, and the birds do not come in contact with the night droppings, which represent the greater part of the evacuation. This dropping board is best placed at a height of 80 centimeters, and can consist of solid, waterproof plywood board. A wooden frame installed under this board will prevent subsequent warping. The surface is

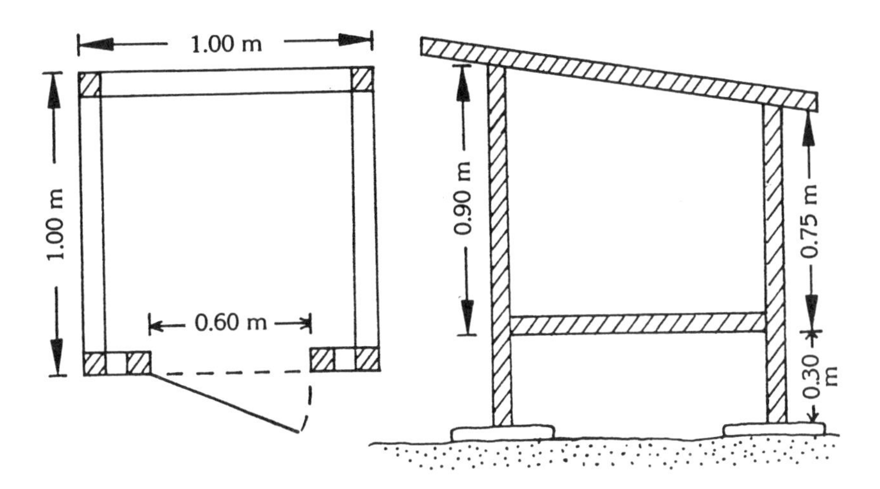

Outline of a small coop suitable for a small flock of bantams or a brood hen with chicks. The door also serves as a window and provides access to the run. The interior should be furnished according to its intended use. If constructed with wider proportions, and then artificially heated, it is suitable for rearing chicks. With a latticed front, it could house young hens.

best provided with three protective coats of coal tar. The perches — consisting of 5 × 5 centimeter thick, planed wooden laths with rounded-off corners on top — should be installed at a distance of at least 50 centimeters from the dropping board. On top of the dropping board, below the perches, it is advisable to place an approximately 15-centimeter-high frame covered with wire mesh, which prevents the birds from coming into contact with the droppings. This will also make the weekly cleaning of the dropping board easier.

For the perches, one should install a supporting fixture — consisting of wooden slats or fairly strong boards with corresponding notches for inserting the perches — on both sides. The furnishings must be constructed in such a way that they can easily be disassembled at any time. In this way cleaning and disinfecting will be made much easier. It is also possible to use angle irons with a metal peg pointing upward for anchoring the perches; corresponding holes will then be drilled in the wooden perches from below. If these openings and the contact points of the perches are treated with oil after the weekly cleaning of the dropping board, the usual hiding places favored by red mites will be uninhabitable.

The necessary laying nests are best purchased complete on the market. A variety of models is offered; the most expensive are not necessarily the best. If it is anticipated that only laying hens will be kept, then one can fall back on the simple family nest. Because of their construction, family nests are relatively dark inside and, for this reason, are very readily sought out by hens. The hens build their own nests in the thick nest bedding of hay or straw, and because of the thick padding there are few cracked eggs.

For successful chicken breeding, trap nests are necessary, at least during the breeding season, in order to determine the parentage of the chicks and thus the hereditary strength of the individual hens for specific traits. Otherwise, only a very small breeding flock with not more than two to three hens for each cock is possible. These trap nests are offered in various models on the market; trap-nest fronts can also be purchased and placed in front of available standard nests. The dimensions of the nests should be 35 to 40 centimeters wide, 40 centimeters deep, and 40 centimeters high; for bantams, 30 × 30 × 30 centimeters is sufficient. So that the hens feel they are hidden, the front of the nest should be narrowed, at first with a 10-centimeter-high board placed on the floor of the nest to prevent the nest bedding from spilling out. For the sides or the top, a narrowing of 5 to 7 centimeters is sufficient. The use of orange crates is not advisable because of the difficulty of keeping them clean. The nests are either equipped with 60–80-centimeter-high feet or hung on the coop wall from sturdy hooks. In the latter case, the back of the nest can remain open, since the coop wall takes over the function of the rear wall.

Landing perches should be installed at a distance of about 20 centimeters in front of the nest; if these can be folded up with the aid of hinges, they can also be used to block the nests. Some chickens have a tendency to spend the night in the nests, particularly when they are placed higher than the perches. This will lead to heavy soiling of the nests with night droppings.

In order to provide the birds with enough room for egg laying and roosting, an average of 20 to 30 centimeters of perch space should be provided for each chicken, and an open nest should be installed for every five hens. (If only five hens are kept, then two nests are better.) When using trap nests, three hens to each nest should be allowed for. Depending on need, the nests can be installed in two to three tiers; in any case, a steep surface should be provided on top, which prevents the hens from flying up and spending the night. A hatch is often built into this diagonal area, in order to use the space above the nests for storing various articles. It is practical, when using trap nests, to always have laying lists, colored leg bands, and so forth, on hand there.

To discourage parasites from staying in the nests, one can use fresh green fern for the bedding — it can also be shriveled or dry, however. On trips to the country it can be collected along the edges of woods; parasites are driven away by the slightly acrid smell. In other cases, some poultry insecticide powder should be added on the bottom of the nest under the bedding, particularly in the warmer months.

The food and drinking containers must be easy to clean, sturdy, and easy for the birds to reach at all times. A box equipped with wire mesh or duckboard, which prevents the litter from becoming wet, is best placed under the watering place. This box should be high enough that the water is not soiled by the litter when the hens are scratching. The diameter must be chosen so that the hens can jump up and perch there while drinking. As a rule, it is better to use plastic or ceramic containers for drinking water (especially when medications or vitamin preparations are added to the water), since metal containers easily oxidize, which could lead to the formation of toxic compounds. Covered drinking troughs are recommended over open water containers. These can be found on the market in various sizes and styles.

Food containers of galvanized metal can always be purchased on the market as long troughs of all sizes (for all stages of development), as round troughs suspended with chains from the ceiling, and as automatic feeders in various forms and sizes. With a little mechanical skill, you can also build wooden troughs yourself; however, you must consider the importance of ease of cleaning. It's better to offer several small troughs! These are easier to handle. The choice of the food container depends on the number of birds and method of chicken keeping. For a small flock of up to ten laying hens, one can get by with one feed trough having a length of 50 to 70 centimeters. The width should be 20 centimeters and the depth about 15 centimeters. So that as little feed as possible is thrown from the trough by the hens, the side walls should be crimped inward. Wire guards spaced seven to eight centimeters along the feed trough prevent the hens from climbing up and scratching out the feed.

For the constant availability of laying mash or a complete ration for laying hens, a round automatic feeder of small capacity is recommended for small flocks. The amount of feed dispensed can be adjusted with most automatic feeders. A narrow setting is recommended for laying mash, since the

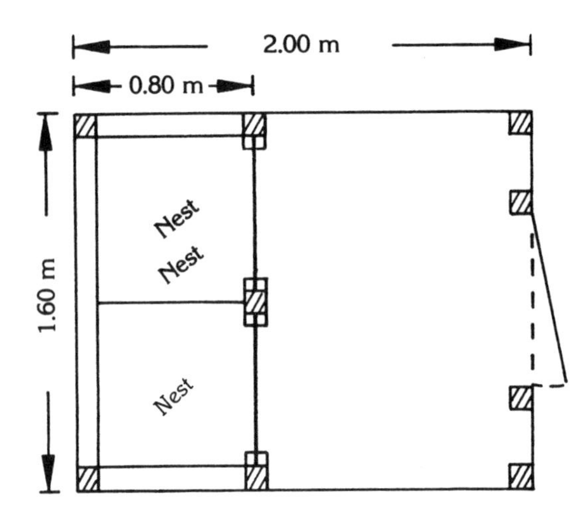

Coop for waterfowl or turkeys. This floor plan accommodates one gander and two geese. The coop is single-walled; it should be double-walled only around the nest area. For turkeys, a perch must be installed at an appropriate height, and the nests—as with ducks—are placed on the floor of the coop.

hens otherwise readily waste the feed with their bills. The grain ration is then thrown in the litter or in a narrow trough in late afternoon or evening. Light breeds get about 50 grams; for heavy breeds, only about 40 grams per chicken is necessary.

For rearing chicks or youngsters, troughs should be chosen that correspond to the age and the size. Bantams, even when fully grown, get by with trough sizes suitable for young hens. Drinking containers are needed in only two sizes: one with a small rim serves for the small chicks, and later the same containers can be used for adult hens. The reasons for raising the chicks and youngsters completely separate from the adults (with natural brooding the broody hen should also be free of internal parasites if possible) will be given later. For this reason, the feed and drinking containers should also be used separately, or must be thoroughly cleaned before switching them.

Geese and Ducks

Geese and ducks make no great demands with respect to housing; they are quite satisfied with a single-sided wooden coop. Nevertheless, this must be draft-free and have an insulated floor to counter cold and dampness. For keeping a few ducks, a wooden shelter of small size can be used. To simplify the necessary chores, the roof should be removable or a section of it should fold up. Sufficient ventilation is necessary even in winter. Similarly, the windows, which need be only half the size of those in a chicken coop of equivalent area, as a rule only have wire mesh in the summer.

Elevation of the coop for waterfowl or turkeys. No glassed window is required; light enters through the grating in the top half of the door, which also provides ventilation. These poultry species are not sensitive to cold. Dry litter must be provided.

During planning, depending on the size of the duck breed that will be kept, ¼ to ⅓ cubic meters should be allowed per bird. To prevent the entry of vermin, the floor should be built of stones cemented together (upon a layer of gravel for insulation) or from concrete with wire mesh. A layer of broken glass makes it more difficult for rats to nest under the floor. When keeping a greater number of ducks or when keeping geese, the coop should be high enough that one can stand comfortably.

For breeding, the only furnishings necessary are nest boxes placed on the floor. These should, if possible, be placed in semi-darkness at the back of the coop, and can be open on top. It is important that the partitions are high enough that geese or ducks which are in the nests cannot see and disturb one another. The area of the nest depends on the size of the breed, but should be at least 80 × 80 centimeters for geese and about 50 × 50 centimeters for ducks. In a coop for breeding geese it is best to install the nest compartments permanently, since the geese prefer to retreat into their compartment and defend their territories during fights. For this reason a nest must be provided for each breeding goose. Ducks (with the exception of

Muscovies and domestic Mallards) are less fussy in this respect; all the same, sufficient laying nests should be available to them. Ducks usually set about egg laying daily in the morning hours, and for this reason should be confined to the coop until approximately 10 o'clock during the laying season. For waterfowl, the nests must be provided with a particularly thick bedding of straw, hay, and similar material, so that no eggs can break, and the nest building instinct, which is present especially in geese and the broody duck breeds, will be satisfied.

If geese and ducks are only going to be kept seasonally until slaughtering size, then no coop furnishings are required. Feed and water are best provided in the run. Make sure that there is enough distance between the feed trough and water container, particularly with ducks. Ducks have the tendency to run to the water with a bill full of feed and take in both together, and in the process waste much of the expensive concentrated feed. A wide spacing between feed and water therefore helps to save feed.

Goose breeding or keeping requires a meadow; ducks, however, can be kept without a grass run; in this case some tender greenfood (lettuce scraps, for example) should be provided to maintain health. The necessary attention must also be given to the care of the run; unless a stream or pond is present, the water dish should be placed in a different spot daily. The water container should always permit the waterfowl to completely submerge their bills. For this reason, simple dishes — approximately 10 to 15 centimeters deep with a diameter of about 40 to 50 centimeters — are best for the drinking water. Cleaning is very quickly accomplished in this case, and if the dish is located near the fence, fresh water can be added from outside with a watering can.

Turkeys and Guinea Fowl

Adult turkeys and guinea fowl are not affected by low temperatures as long as the coop is dry and absolutely free of drafts. The same coops which were recommended for chickens can be used for housing them. The interior furnishings will have to differ in several points. For instance, the perches, corresponding to the larger size of turkeys, should have a diameter of about 8 to 10 centimeters and should be placed about 70 centimeters apart. The nests should be placed at a low height and should act as hiding places for the laying turkeys. This is accomplished by having the entrance hole of the nest, which should have the dimensions of about 60 × 70 centimeters, only 40 centimeters high. The height of the back of the nest should, however, be about 80 centimeters, so the top of the nest should slope toward the front.

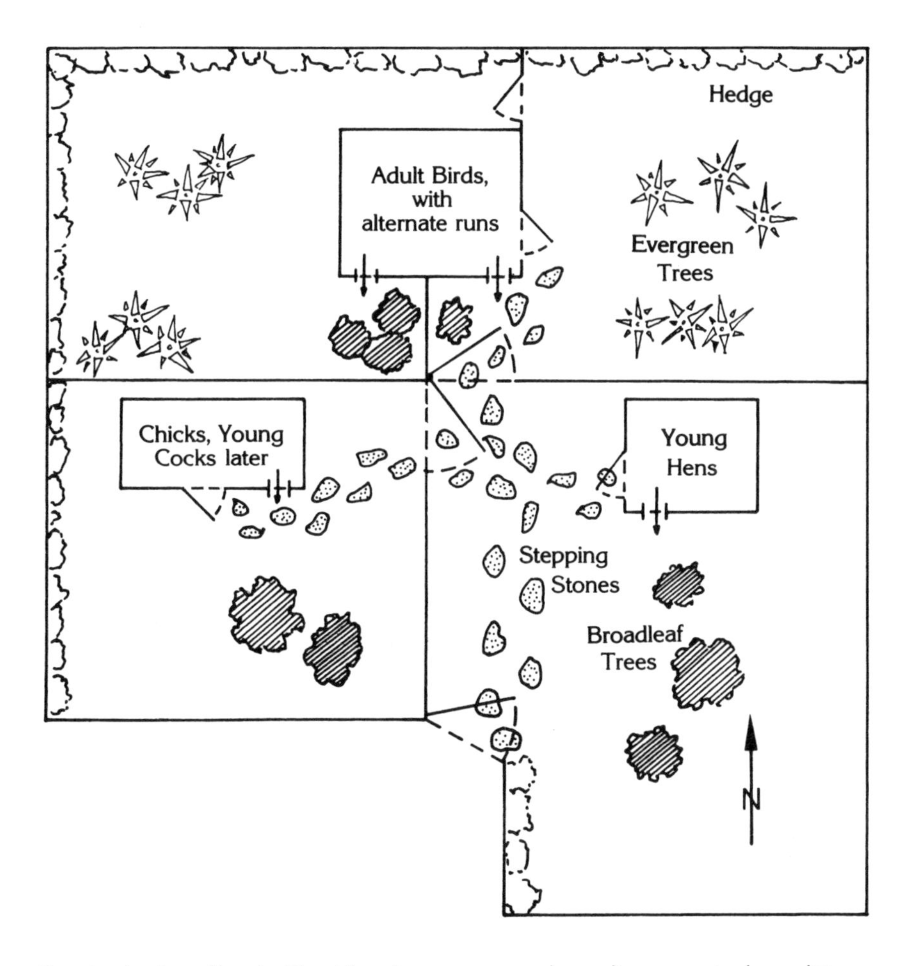

Sketch of a breeding facility. The size may vary, depending on actual conditions; the runs for chicks and young hens can also be used for other poultry. The kinds of trees, and also the hedge, may be chosen according to preference; however, a mixture of broadleaf and evergreen trees is advisable. Evergreens protect against wind in the winter, while broadleaf trees allow maximum sunlight in the winter but provide the desired shade in the summer.

Guinea fowl use the same kinds of nests as chickens; the difference is that they should be placed close to the floor. If the nests are placed in the quietest part of the coop in semi-darkness, the guinea fowl, which are fussy in this respect, will accept them.

For feed and drinking water, the same troughs and drinking containers recommended for chickens can be used. Since turkeys, in particular, require an extraordinary amount of fresh air, a portion of the window should be replaced with wire mesh. The particular requirements for rearing turkey chicks are given in the section on rearing.

The Run and Its Construction

This is the element that distinguishes poultry keeping by the hobbyist, who besides profit also derives a financially unmeasurable value of pleasure and moral uplift from the birds, from the production plants of commercial poultry farms, which are an industry. There the natural requirements for exercise and appropriate accommodations are unconditionally sacrificed for the sake of profit. In the planning and construction of the poultry run, we should always keep the birds' well-being uppermost.

Chickens and Bantams

In general, a grass run with trees and bushes is the optimal living space for chickens. In this case — depending on the available plot of land — various opportunities present themselves. If an old orchard is available, then we already have almost ideal conditions for a poultry run. To minimize damaging ground winds, hedges of evergreen shrubs or densely growing deciduous bushes should be planted. Of course, right from the start limitations are placed on the available area far too quickly. If one is able to rouse the entire family to enthusiasm for poultry keeping, then perhaps a large part of the decorative garden — containing robust plants — including the areas of lawn that have to be mowed regularly, can be offered to the birds as a run. One can keep the run out of sight by planting low bushes and perennial flowers in front of the fence. A simple wooden coop — preferably painted an inconspicuous green (window, fittings, and the edge of roof set off in white) — does not make a disturbing impression in a naturally planted garden. A flock that is not too large — perhaps one cock and six to eight hens — can always be the focal point of this part of the garden. The great diversity of form and color found in poultry breeds offers something for every taste. Accommodations of this kind have yet to be given enough consid-

Above: A lawn run—whether natural or artificially landscaped—offers youngsters the best opportunities for growth.

Below: Every keeper or breeder must recognize that he has assumed responsibility for living creatures.

eration. Thus, living animals can certainly be included as an stimulating element in the creation of a garden.

If the run is expected to continuously provide the chickens with natural food, the lawn should be kept short and fresh through regular mowing or grazing by sheep or geese; overripe grass is not accepted by chickens. When necessary, a balancing fertilizer should be applied. The hens' droppings contain abundant amounts of nitrogen, so that the supplementary fertilization need only consist of potash, phosphate, and calcium. It is beneficial to apply these mineral fertilizers in late winter. Phosphate and potash are best bought as a prepared mixture; this should have approximately the proportion of one part phosphoric acid to two parts potash. Depending on nutrient content, 200 to 300 grams is spread per square meter. For maintaining the fertility of the soil, a sufficient calcium content is necessary. Calcium combines with the acids that occur or are created in the soil and increases the activity of the microorganisms so important for the life of the soil. The pathogens expelled in the birds' droppings are rendered harmless through the combined action of these soil organisms, plants, and weathering influences. In heavy soils (that is, soils with a high clay content), the calcium should be given in the form of lime (this also has a disinfecting effect), and in light, sandy soils in carbonate form. At intervals of no more than two to three years, calcium amounting to one kilogram per square meter should be applied; however, it is better if 300 to 500 grams of it are spread annually in late winter.

If there are too many birds in the run then even the best of care is useless. The grasses are uprooted by the birds and can no longer process the droppings and the pathogens they contain. For a harmonious coexistence, each bird should be provided with at least the following minimum amount of space: breeder chickens should receive 15 square meters, while pure laying hens get by with 10 square meters. For chicks, at least two square meters should be allowed initially, which should be increased to 8 to 12 square meters in the course of development. If only a limited run is available for the brood, then great significance should be attached to the regular selection of the birds. The surplus young roosters are therefore best raised to slaughtering size separately without a run. The same procedure should be used with stunted hens or those with anatomical defects, as well as faults in the characteristics of the breed. If there are still too many birds for the run's capacity, then the surplus should be sold as soon as possible.

An additional alternative is offered by providing alternate runs; in this case the available run is divided in half with a fence, which does not have to be as high as the outer fence. One run can then recover during the weekly rotation, and can handle up to 25% more birds. The pen used for rearing should subsequently remain unused or should be occupied by different species of birds. Only then can a healthy brood be expected. We need only turn to the comparison with free-living wild poultry: the hen leads her chicks away from the flock, and in so doing reduces the danger of possible infection of the youngsters to a minimum.

In order to provide the birds in the run with a place of safety, a number of bushes should be planted. Robust varieties, which can withstand the chickens' scratching around the roots, are especially suitable. The plants should be adapted to the region and soil; in fertile soils, native elder or other berry bushes are good choices. With the latter, one will hardly be able to harvest the berries one's self (the chickens will see to that), but in this case the birds' welfare is, after all, of primary importance. Recommended are currants — especially the black ones — and raspberries. After planting, the hens must be kept away from the root area for at least a year; this is best accomplished with chicken wire and simple posts. Once the plants are well rooted, this protective fence can be removed. In the event that small rocks are available in the area, then these should be laid out in the planting area, in order to hold the hens' scratching activity under the bushes within limits. A small group of about three conifers — possibly spruces — for an evergreen hiding place is also advisable if the run is large enough.

Small runs should be kept free of plantings, however; a few diagonally placed wooden boards or reed mats serve to break up the monotony of the run, and thus increase the birds' well-being. In bright sunshine the areas of shade produced by these stands are readily sought out. In summer, fresh water can also be offered there. Unless the side of the run open to the weather is protected by buildings or existing plantings, a hedge should also be planted. Deciduous hedges have the advantage of allowing more light through in winter, when there is otherwise little light. The available space is always the determining factor in the selection of the poultry breed to be kept; if little space is available, then one of the bantam breeds should be used. Among these one finds stocks that can easily rival the productivity of many large chicken breeds. In addition, the basic requirements of protection from wind and adequate sunshine, as well as sufficient areas of shade and perfect water drainage, must be taken into consideration. Through an attractive planting, everything will be suited to the area:

Above: With geese, opportunity to bathe is necessary for good fertility.

Facing page, above: A pair of Mallards feeding. Aquatic vegetation constitutes a major portion of their natural diet. **Below:** A group of Muscovy drakes and ducks in a pool.

trees and shrubs that are adapted to the habitat (that is, the predominant species growing in the area) should be used. In the breeding and keeping breeds with light-sensitive plumage colors, such as the buff-colored varieties, more shade must be provided in the runs in order to keep the feathers from bleaching. For this purpose, corn or raspberry bushes can be planted in part of the run, among which the birds readily stay. If their basic requirements are attended to, then no limits are placed on the imagination in the creation of the runs. Poisonous plants such as yews should, of course, be kept out of the birds' reach.

Geese and Ducks

A meadow is the basic requirement for keeping geese. A swampy meadow, however, is not suitable for this purpose: first of all, the sour grasses that grow there will be rejected by the birds — even though they are waterfowl — and second, coccidia and maw worms spread particularly rapidly in standing water. For this reason, sufficient water drainage must be ensured. Up to an age of three months, the youngsters must in any case be confined to dry land. For fertilization, the comments previously made in connection with chicken runs apply. For geese, the fence need only be one meter high; for ducks as little as 60 centimeters will do, depending on the breed. The duck breeds capable of flight are an exception.

Waterfowl always require — and with alternated meadows in each meadow area — sufficient shade. If there is no possibility of planting trees and bushes, then shade should be provided by setting up rush mats. Although a small, clean body of water is not a requirement for keeping a few ducks or geese, it is an absolute necessity for successfully breeding most kinds. A stream bordering the plot of land would certainly be a decisive advantage for breeding waterfowl, but one can also achieve good results with a small man-made pond. Unless one wishes to build a basin of concrete reinforced with wire mesh, one can turn to the prefabricated pools that are offered on the market for building aquatic gardens. In this instance one should make certain that an area with a gentle slope is present for climbing in or out comfortably. Ponds lined with metal foil do not usually last very long, since the birds will hasten the wear and tear with their bills and toenails. Quite satisfactory results have also been achieved with the use of a plastic tub, such as are offered in various stores as sand boxes for small children. These are usually square with a side length of 100 to 120 centimeters and a depth of approximately 20 to 30 centimeters. Regular cleaning of the water basin is very important. In order to simplify this chore, the deepest point of the basin should be equipped with a drain. Depending on the ground and water conditions, a seepage shaft or a drainage pipe to the nearest sewer should be provided underneath.

Turkeys and Guinea Fowl

For turkeys, and particularly for guinea fowl, free run would be ideal. This possibility would, however, be available only in the rarest instances. For this reason, the choice of poultry should be governed by the available space, and the number of birds should also be reduced if necessary. For turkeys, a run with a meadow and stands of trees (beeches and oaks) is the most natural. A run area of 50 square meters should be allowed for each turkey to be raised; for a breeding flock, as much as 100 square meters are needed for each bird. The fence must be two meters high for turkeys; for guinea fowl, which require approximately 30 square meters of run per bird, the run should also enclosed, if damage to or disturbance of the neighbors' property are a possibility. Turkeys and guinea fowl require the same run conditions as chickens. Constantly wet areas are not suitable for these poultry species.

Building the Run

Construction of the run begins by leveling the areas on which the fence will stand. First the corner posts with braces are inserted. These posts are then tied to one another with cords or strings; the string stretched close to the ground is used to align the posts that will be inserted at intervals of 2 to 2½ meters. An additional string placed higher up serves to line up the posts. Depending on the height of the fence to be erected, the posts should be buried 50 to 70 centimeters deep. If wooden posts are used, they must be given a durable protective coat unless they are pressure-treated. This is unnecessary when using concrete posts. Today, metal posts are usually coated with plastic, but they should be equipped with a concrete footing for the sake of better stability. It is easiest to buy or rent a posthole-digger — there are also such tools for manual operation — and then dig the holes in the locations provided. These holes are then filled with fresh concrete, and, finally, the metal posts are pushed into the still soft mass.

For covering the fence, galvanized or plastic-coated wire netting is best. If the post is two meters high, in addition to the upper and lower guy wires another should also be strung in the middle. For adult chickens, a mesh size of 70 millimeters is adequate; for youngsters and chicks, however, it should not exceed 50 millimeters. For chicks, the lower part of the fence should either be blocked with very fine mesh — so-called chick mesh — or, better yet, with wood. In addition, prefabricated fences in various styles are offered on the market today; a careful price comparison is by all means recommended.

Depending on the size of the run, the gates in the fence should be built sufficiently wide. Normal gates should be one meter wide, so that unimpeded entry for necessary maintenance, with a wheelbarrow as well, is ensured. Should the entry of a motor vehicle occasionally be necessary, then it is best to build a removable area of fence about three meters wide in the appropriate location.

The fence height must correspond to the breed being kept. Light chicken breeds always require a fence 2 meters high; most medium-weight breeds get by with 1½ meters, and the heavy breeds with a 1-meter-high mesh fence. For the few bantams that can fly very well, the complete enclosure of the pen with a covering of wire mesh or netting is necessary.

Below: A very close relationship can develop between poultry and keeper.

Facing page: As pets, chickens are enjoyable and easy to care for.

WHICH POULTRY FOR WHICH PURPOSE?

Chickens

In order to provide some assistance in the task of choosing the appropriate chicken breed, the breeds' specific characteristics are described here. If little value is placed on unusual appearance or beautiful coloration, then the hybrid chickens, which are bred for high production, are recommended; however, these birds, which are available from any poultry breeder, are the end products of a series of crosses. The characteristics of their offspring will vary greatly. For the small-scale poultry keeper, the very dependable medium-weight, dual-purpose hybrid chickens are highly recommended. These hens lay more or less brown-colored eggs in large numbers and will maintain such a level of production for two years. After that, they are the basis of all manner of fine things for the kitchen. The light, white hybrid hens are less suited for keeping in small flocks because of their excessive nervousness.

Mediterranean Breeds

All light breeds that are still vigorous are suited for egg production. To this group belong all the breeds from the Spanish-Italian region — the Mediterranean breeds — as well as the old indigenous chicken (landfowl) breeds of Germany. Let us begin with one of the oldest Spanish chicken breeds, the Castilian Fowl. Poultry breeders who are also Castilian fanciers praise the diligent laying and the size of the eggs of this breed, which is bred only in the black color variety. The feet, beak, and eyes of these lightweight chickens are a dark, almost black color; the white ear-lobes on the sides of the head and the bright red head points (comb and wattles) form an attractive contrast to the black plumage with its greenish sheen. (By "ear-lobe" is meant the skin tissue occurring in the ear region, as long as it is white; if, on the other hand, it is red then it is an "ear-wattle.") The birds have a stately, somewhat upright carriage, and draw up the tail feathers to an angle of 90 degrees to the line of the back.

The oldest of the chicken breeds that have apparently been bred on foundation of the Castilian is the Spanish chicken. The most prominent characteristic of this breed is the white covering of the entire facial area. This white face forms a charming contrast to the red comb and wattles as well as the lustrous green black plumage. This breed also occurs only in the black color variety. Because of the small breeding stock remaining today, the laying production is no longer as high as it once was; the large egg, however, is firmly anchored in the genotype, so that with appropriate selective breeding the production could be raised again. This breed, because of its large head points, is relatively sensitive to frost, and should be housed with this in mind. (By "head points" are meant the comb, ear-lobes or ear-wattles, and wattles.)

The Minorca is also descended from the same genotype. In the same way as its white-faced cousins, it was also perfected as a breed in England. Unfortunately, in the process, the tremendous production potential was somewhat neglected at the expense of larger, white ear-lobes and the other head points. (This fate was already suffered by several once fashionable breeds.) Minorcas are upright, large-framed chickens with an elegant carriage. With this breed, besides the black color variety with black feet and dark eyes, white ones with flesh-colored-white feet and red eyes are also permitted. In addition, besides the birds with a single comb, there are also those with more cold-resistant rose combs. A rose comb is a flat, fleshy structure with small pearl-like skin protuberances that sits on top of the head.

The Andalusian also originated from the Castilian. This breed is always presented as a classic example of intermediate color inheritance, and should be as familiar to almost everybody in biology class as Gregor Mendel's red and white flowers. In those days the color of these chickens, Andalusian blue, was unique. The ground color is a variable pigeon-blue in tone with dark feather edging. The rooster's plumes are darker and appear almost black. From blue parents, half of the offspring are blue, and a quarter each are either black or splashed-white. If one mates the black and white birds with each another, then one obtains only blue offspring, some of which, however, can be a lighter or a darker shade. The birds of this breed are not heavier than the ancestral breed, the Castilian, but they do have a more elegant stance with a low tail carriage. The ear-lobes, as with all Mediterranean races, are white; the eye and leg color is dark. Birds of vigorous stock show good laying production.

The Italian and the Leghorn make up the second group of Mediterranean breeds. Both breeds have indigenous Italian chickens as ancestors, and are known to be very good laying chickens. Common characteristics are, besides the elongated, elegant form, the yellow leg color and (as is true of all Mediterranean breeds) white ear-lobes and white

Above: The Red Junglefowl is the principal ancestor of domestic chickens.

Below: The Grey Junglefowl, *Gallus sonnerati.*

eggshells. While Italians were developed as a breed in Europe, the Leghorn became the number-one production chicken in the United States. Today, the foundation of commercial poultry breeding is the Leghorn and the laying hybrids that were bred from it.

The Leghorn only occurs in the white color variety in Europe. In the show cage, it shows an elongated shape with delicate bones and head points, as well as a lush tail area. The partridge-colored variety of the Italian once had the widest distribution of any high-production, run-kept, laying hen. Only with the advent of the practice of keeping chickens in coops the year round were they displaced by the Leghorn. The partridge-colored rooster is what every layman supposes the colorful lord of the barnyard chicken flock to look like. Pedigree poultry breeders have bred almost all possible color varieties of chickens from this breed, and, in addition to the original single-combed, the rose-combed as well. The fancier of colorful poultry has the widest choice with this breed. The production varies, depending on the ancestors used for breeding the diverse color varieties. Moderately productive varieties include the partridge, gold, silver, orange, white, black, and also the sex linked. This variety is called "sex linked" because one can already distinguish between cocks and hens by the chicks' down plum-

An Italian rooster: the single comb and white ear-lobes are characteristic of the breed.

age (which is otherwise impossible).

Sexlinked Leghorns exhibit barred markings on the partridge-colored ground color.

Northwest European Breeds

The next group of laying breeds are the Northwest European breeds, which are indigenous to various regions. Unfortunately, these have been neglected because of the "fashionable breeds"; with careful breeding selection their respectable laying production could also have been improved. They were adapted to their particular regions and climates, and provided very good egg yields with the simplest of care. Because of the keeping and feeding restrictions, the laying period usually did not begin until March, when the optimal light conditions were attained and fresh, protein-rich food could be found. These breeds then laid very diligently without pause into the fall. As is true of almost all laying breeds, these were seldom prone to becoming broody (broody = ready to sit).

Besides the white eggshell, the Northwest European breeds have in common white ear-lobes, long plumage, and usually blue-gray legs. On the basis of their unusual markings, common ancestors are concluded for the "spangled breeds"; for this reason they are also very similar in their useful traits. The breeds are the Ostfrisian Moven, Braekel, Westphalian Totleger, Frisian Fowl, and Hamburg. Hamburgs also have the so-called spangled marking; by this is understood round, black dots with a green sheen, always on the tip of the feather. The spangled, laced, or mottled marking, as well as the band-like marking of the Braekel, always occurs on a silver-white or gold-brown ground color. Hamburg chickens also occur in black, white, and blue. All of these breeds have a more or less elegant, slim landfowl form. These breeds reach their optimal productivity in grass runs which are as large as possible; an almost unlimited run would be ideal, since they would then find much natural food, which would reduce the cost of keeping them. Closely related to the previously mentioned breeds is the Lakenvelder; the trunk plumage is white, but the neck and tail feathers are almost black. This breed has experienced renewed popularity in recent years, and has become more vigorous.

An additional group of native German breeds is formed by the Mountain Crower, the Mountain Floppy Comb, and the Creeper. If one can believe the traditions, the ancestors of these breeds (the Crower in particular) are said to have been brought back from the Near East during the Crusades. The unusual trait of the Crower was promoted through "crowing contests"; for that purpose the length of the roosters' crow was (and is) judged in the spring. To our knowledge, this trait is otherwise found only in several Japanese chicken breeds. Mountain Crowers thus represent a unique cultural resource which is worth preserving. Floppy Combs presumably also count Spanish chickens among their ancestors and therefore have the large single comb, which in the hens tips over to the side and "flaps" when they run about. Greater egg production and heavier eggs were brought to this breed with the Spanish genotype. For this reason they were once a popular laying chicken in the mountains. Unfortunately, this breed has also been neglected by breeders, and today is numbered among the rarities in poultry breeding. We find a unique type of marking — the so-called edging — in this breed. This marking is the coarsest form of penciling; a golden-brown or white spot is found in the middle of the otherwise black feathers. The neck and tail plumage is black. In the rooster, for genetic reasons, the neck is marked with gold or white in the same way as the saddle feathers.

While only the black-gold-edged color variety is recognized in Crowers, Floppy Combs are found in gold-edged black, white-edged black, black, and barred. In addition to the same color varieties as the Floppy Combs, Crowers are also found in white. Creepers are an old landfowl variety with shortened legs. The short legs originated through a mutation which is lethal in purebred form; purebred Creepers die in the egg. Of the expected chicks, therefore, two thirds will be short-legged and one third will have normal legs. Creepers have standard single combs, an elongated, cylindrical form, and show normal laying production. The laying production varies somewhat in the different color varieties; the short-legged birds are also content with fairly small runs.

The Rheinlander is also a very good laying chicken; it was developed from the robust Eifeler Landfowl. This breed has a small rose comb which is not subject to cold, and an elongated, rectangular body, which provides sufficient room for the organs critical for laying. Although the first color varieties were the black-red and the white, today the blacks are the most widely distributed. But in this case the perfect "show beauty" has been obtained at the cost of production. This can be observed in many breeds; all the same, there are numerous flocks of the black color variety that also show good production. The black-red as well as the newest color variety, the "silver-necked," are said to lay very large eggs. In addition, one also finds blue and barred Rheinlanders.

Another breed with principally laying characteristics is the German Barred Fowl, a slightly more upright indigenous chicken with a single comb (which tips over to the side in

Leghorn chickens are good egg producers.

hens), flesh-colored legs, and the feather markings that give it its name.

The tailless Rumpless Fowl is a very light indigenous chicken with a loss mutation as a characteristic of the breed. This breed also has a single comb and occurs in almost all common chicken color varieties.

The Bearded Thuringer with its prominent beard is another interesting chicken. This hardy laying chicken matures quite early, has a single comb, and is bred in numerous color varieties. Particularly attractive is the spotted variety with golden-brown or silver-white ground color and black spots. In type it is slightly compact but very balanced.

From the Northwest European breeds the spectrum apparently extends seamlessly over the bearded breeds to the crested breeds. These originated regionally in the Middle Ages, and because of their special nature were carried along on the trade routes and by this means were dispersed. Through crosses with the respective indigenous fowl, many other breeds were developed at the time.

Thus the Bearded Appenzell originated from such crosses, the goal being a laying hen suited to the rugged alpine region, which could find most of its own food in the country-

Naked Neck Bantams are striking because of their featherless necks.

side. The flat rose comb offers a very small surface for cold to attack, and the smooth, undivided wattle protects the throat and face. This chicken, which only occurs in the black color variety, has white ear-lobes (which, however, are covered by the wattles) and lays white eggs.

Crested Fowl and Their Relatives

Substantially lighter and more lively are the Crested Appenzells. A particularly attractive feature of this breed is the pointed crest; this consists of straight, erect feathers, which grow on a cranial process. In this breed the comb consists of only two small round horns. Fine, medium-size wattles and white ear-lobes complete the head points of this breed. The accepted colors are very attractive; besides black, silver-black-spotted and gold-black-spotted varieties also occur. In this breed the black spots are not round, but are more crescent-shaped. Unless this breed can be provided with an unconfined run, they should be given a pen that is also enclosed on top. In an unconfined run they will forage for a great deal of food themselves and they will develop their maximum production, which, nevertheless, cannot compare with the most productive breeds with respect to egg size. In winter, as is true of many landfowl varieties, one can count on only minimal production with normal accommodations.

One of the oldest crested-fowl breeds is the Polish. The form of this breed corresponds to that of the light landfowl; Polish have the most complete crests, which rise from a bony cranial process, the so-called knob. In addition, all crested fowl have higher, enlarged nostrils (connected with the structure of the cranial process). The crest must be proportionally large, and should be compact, thick, and full. Breeds with full crests should be kept in covered runs in wet weather. Unless they will be exhibited, it is advisable to trim the sides of the crests with scissors so the chickens can see better. Because of the feather structure, the rooster's crest is thinner and looser. Some Polish have beards and are bred in the solid colors black, white, and blue, as well as silver-white and gold-brown with black feather edging; nor should the leather-yellow with cream-white edging (chamois) be overlooked. Still more attractive — because of their white crests — are the beardless Polish; these are bred in black, white, blue, mottled, and barred. Although today egg production is secondary to ornamental value, it must still be designated as very respectable. Crested fowl in particular have been immortalized in the paintings of old Dutch masters.

The French crested breeds are the Houdan and the Crevecoeur; they are primarily raised for meat, and have a long, broad barnyard-fowl shape with a full breast. Both breeds are bearded; the Crevecoeur have (inherited from the Polish) a horn comb, or, if the horns branch out, a V-shaped comb. In the Houdan, the characteristic of the breed is the V-shaped comb, made up of two flat sections, and the fifth toe pointing to the rear (inherited from the Dorking). They occur in mottled, black, white, and pearl-gray; the standard color varieties of the Crevecoeur are black, blue, white, pearl-gray, and barred.

Sultans, which are said to come from Eastern Europe or the Orient, should be classified between large fowl and bantams on the basis of their appearance. They have a short, stocky barnyard-fowl shape, a barely medium-high carriage, and, besides the full, round crest, a full beard and well-feathered legs with long leg plumage. This five-toed breed is only bred in white, and is a purely ornamental chicken with cultural significance. Recently it has again turned up more often at poultry shows, after one feared for the existence of this breed a few years ago.

The Brabanter — a breed already known in the Middle Ages — unlike the previously mentioned breed, should not have a full crest, but should instead carry a helmet-shaped crest with erect feathers. In Germany only the Dutch breeding strain is accepted, although a Belgian strain also exists. Besides black, blue, pearl-gray, white, and barred, the color varieties silver-white and gold-brown with black, crescent-shaped markings on the tips of the feathers, as well as chamois, are very popular. This light land-fowl type of breed is a good layer without exception. The comb, which consists of two horns, as well as the beard, which is divided into three parts, are additional characteristics of the breed.

As odd as it may sound, there are crested fowl without crests. The Bearded Owls, closely related to the Brabanter, have no crest even though they have the enlarged nostrils. As with the Brabanter, the comb is made up of two spikes. The full beard, which gives them their name, should cover the face and throat like a veil, and for shows may not display any separation, as is required in the Brabanter. Productivity and color varieties are as with Brabanters; the only difference is that the height is somewhat less.

Another Dutch breed belonging to the crestless crested fowls is the Breda. This relatively upright-standing chicken is characterized by feathering of the feet and very long thigh feathers, the "boots." All that remains of the comb is a small depression with a thickened rim covered with red skin. The crest can only be distinguished by the brush-like plumage behind the sunken comb. The ear-lobes are small and white. The medium-length body has a high, full breast which indicates a good broiler in addition to normal egg production. According to reports from a number of breeders, this breed displays a relatively dependable brooding instinct. Today one can find the Breda at shows — especially at the special shows for rare chicken breeds — in the colors black, blue, pearl-gray, white, and barred.

The La Fleche is an old French breed. The head ornamentation of this breed, which is also somewhat more upright, consists of parallel spikes which should stand as close to vertical as possible over the eyes. The crest can only be distinguished by the brushlike tuft of feathers behind the comb. Besides very respectable egg output, this breed, which principally occurs in the black color variety, is also credited with good meat production. Therefore, a slightly erect body which is broad, full-bodied, and long is called for. The ear-lobes are white; apart from black, there are also white, blue, and barred La Fleches.

Also related to the crested fowl are the Augsburg Fowl. This breed was bred in Augsburg and the surrounding countryside as well as in the Black Forest. The unique char-

The imposing head of a Malay rooster is representative of the breeds of the game group.

acteristic of the breed — the buttercup comb — developed from the combination of the V-shaped and single combs. This medium weight and also medium-high-standing chicken only occurs in the black color variety; the white ear-lobes and the red comb form an attractive contrast to the color of the plumage. From the ancestral races, the black Italian and black La Fleche, this breed has inherited the genes for good egg production and very respectable meat yield.

Multipurpose Breeds

Saxon Fowl are also one of the breeds that produce large numbers of eggs and an adequate amount of meat. Currently this breed is one of the rarities at West German poultry shows. Various breeds have contributed to this breed, presumably including the Minorca and also the Langshan. From the latter the tinted to light-brown eggshell is apparently derived, which is required in spite of the white ear-lobes. A delicate single comb adorns the head; because of the unusually abundant saddle plumage, the rising line of the back extends without interruption into the full but only medium-length tail. The leg color is blackish-gray with dark eyes in the black color variety; in the white and barred varieties the legs are light flesh-colored and the eyes are red. Importations from East Germany give hope for the revival of the stock.

Above: The Araucana is remarkable because of its prominent head feathering.

Below: The absence of a tail is another feature of the Araucana, which originated in South America.

The Ramelsloher, which originated in the vicinity of Hamburg, were bred about 1870 from chickens suited to "indoor chick production" (winter layers, whose chicks were fattened in heated rooms and could be sold to Hamburg's well-to-do inhabitants at Easter), the off-colored Andalusian, and presumably also the buff Cochin. These somewhat upright chickens should have, besides the elongated, cylindrical body, a full breast and a standard single comb. The ear-lobes are bluish-white, the eye color is dark, and the bill and legs are a shade of blue. They are also seldom seen at poultry shows — always with white plumage, even though a buff color variety is also allowed.

Indigenous to the same region are the Vorwerk Fowl, for which, besides buff Ramelslohers and Lakenvelders with their black neck and tail plumage, buff chickens of Asian origin were possibly also used to develop the breed at the beginning of the century. The yellow eggshell color as well as the white ear-lobes, which are often edged with red, strengthen this supposition. This robust chicken with its compact form is credited with rather good production. The black neck and tail plumage with its slight deposits of buff looks quite charming together with the buff body plumage. The small, single comb may tip slightly toward the rear in the hen. Red eye color and blue-gray leg color are required.

The old landfowl breeds Altsteirer and Sulmtaler originated in the Steiermark (a province in East Prussia). Both breeds at one time had great commercial importance and were purposely bred for laying. The lighter Altsteirer are primarily laying chickens and are recognized in brown (similar to the black-red color) as well as white color varieties. Sulmtalers are more compact, deeper bodied, and heavier; they are bred for the best fattening ability and are only found in the wheaten color. Characteristic of both breeds is the feather tuft on the back of the occiput, which consists of only a small clump of feathers in the rooster because of the sexually determined feather structure. The hen's comb is pushed to the front because of the more strongly developed tuft, forming a cushion-comb. These breeds have a common-landfowl shape, flesh-colored to white legs, and red eyes; the ear-lobes are must be white, but not infrequently have a reddish rim. In these breeds the white color of the ear-lobes often spreads to the red facial area. This serious fault is, however, only of significance in show birds. The Altsteirer's eggs are ivory-colored, while those of the Sulmtaler are cream-colored to light brown.

The Dorking, an interesting breed from England, was already present there in pre-Christian times. Dorkings have handed down their traits to many breeds. Besides the long and heavy body, the fifth toe, which points freely to the rear over the fourth toe, is just as characteristic of the breed as the abundant, long plumage. Dorkings possess relatively good meat-chicken traits; according to reports from a number of breeders, the laying results are also supposed to be satisfactory. The cultural significance of this breed should, however, be of prime importance for preservation and purebred breeding. Dorkings have, depending on the color variety, either a large, single comb or a rose comb; the color varieties silver-gray with single comb, dark with single or rose comb, white and barred with rose comb, and red with single comb are allowed. In the single-combed hens, the combs are carried tipped to the side. The legs of these chickens are flesh-colored, the eyes are orange-red.

Dominiques come from the United States, but nothing else is known about how they were developed. This breed, the oldest from North America, has a somewhat elongated landfowl form and is moderately upright. Only barred Dominiques with yellow leg color and fine rose combs are found. The ear-lobes are red, the eyes orange-red; the eggshell color must be brownish. This breed played a large role in the development of Wyandottes. Dominiques can make very good use of a grass run, since they nimbly search in all nooks and crannies for natural food.

The story of the development of the Naked Neck is worthy of notice. They obtained their naked neck as a result of a loss mutation, since when they are crossed with normally feathered chickens, the bare neck, except for a feather tuft in the middle of the front side of the neck, is always passed on. These chickens, which are of common type, carry their elongated, cylindrical bodies slightly raised; not only are the visible naked neck and crop area, which show the red skin, unfeathered, but the areas of skin that are covered with down feathers in other chickens are also bare; these areas are covered by the remaining plumage. The robust bare necks were developed in the breed in Siebenbürgen and in Germany; they tolerate dry cold very well, and are known to be good laying chickens. As is true of all chickens, damp cold does not suit them. Although the rose comb is also allowable, birds shown at exhibitions usually have a small, upright single comb; the ear-lobes must be red, the eyes orange-red. The blacks, which are exhibited most often, as well as the blues, have a blackish-gray to slate-blue leg color; the color varieties white, barred, red, and buff have flesh-colored legs. Besides an impressive number of heavy eggs, Naked Necks are also credited with provid-

The Malay Game hen shows a resemblance to a bird of prey.

ing a tasty roast, as the meat is said to be very short fibered.

Another curiosity is the Araucana, which comes from South America. Presumably they are descended from chickens kept by the inhabitants of the South Sea Islands, who reached and settled the western coast of South America (present-day Chile and Peru) with the aid of rafts. For reasons unknown to us (possibly malnutrition, displacement by wars, volcanic eruptions), the original inhabitants of the South Sea Islands may have been forced to leave their homelands. Live chickens were certainly taken along as a food supply on the journey into the unknown. Apparently because of the limited stock available, more intensive inbreeding took place than with our domestic chickens, which resulted in a unique mutation, the light-greenish-blue eggshell color. Of course, this hypothesis cannot be scientifically substantiated. The dominance of this mutation has, nevertheless, been proven, and it was carried to all of the various forms of this breed in its area of origin. After the conquest of South America by the Spaniards, chickens from the Mediterranean region were certainly crossed with the native chickens; it can also be assumed that in the recent past "more modern" chicken breeds reached the region of origin of the Araucanas via trade routes and were crossed in. The mutation for the blue eggshell

Yamato Games are among the smallest of the non-bantam breeds.

color (which, in contrast to the brown, colors the entire calcium shell) is commonly thought to be passed on to the offspring invariably. This chicken breed was first discovered among an Indian tribe in Chile in 1880. Besides birds with "ear-tufts" (these are skin appendages covered with feathers on or in the vicinity of the ear-lobes) some birds also have a beard, some have both the ear-tufts and the beard, and many others have the mutation for taillessness, already familiar to us with the Rumpless Fowl. In addition, the colors of almost all common chickens were present. The common characteristic of the breed was and is the light-blue or greenish-blue to turquoise-colored egg. In order to arrive at a uniform appearance for exhibition purposes, the show standard of the purebred breeders in West Germany recognizes only the tailless type as an Araucana. The birds should have ear-tufts and a beard, or both; the ear-tufts usually are not equally developed. The comb is an irregular pea comb; if a beard is present, no wattles are usually visible. The leg color should be willow-green to greenish-yellow in most color varieties, and olive-black if the plumage is predominantly black. The eyes are red to orange-red. This fowl, which is on the light side, is very docile and a good layer despite its lively nature. The chicks are often brooded and attentively reared by the hens themselves. For the poultryman who has no interest in purebreds, birds with tails can also be used; if one only keeps these interesting chickens for the sake of the eggs, then one can often purchase the tailed birds (that is, possessing tails) that become available cheaply from exhibition breeders. Breeders' addresses can be found in the trade journals, poultry press, and other magazines; in addition, opportunities to purchase special chickens are offered through specialized societies or at poultry shows.

Games and Related Breeds

Probably the oldest breeds of chickens are the game, or fighting, breeds from the native lands of the Red Junglefowl. It can be assumed that the first domesticated chickens, besides their still minor value for feeding people, also served cultural purposes and as sacrifices to the gods. In this connection the rooster's fighting instinct surely played a special role. In this way, over the course of thousands of years, the various game breeds and varieties were developed.

The Aseel, with its various regional varieties from India, and the game of upright Malay type, are among the oldest game breeds. The Malays exhibited in our shows received their extremely erect carriage with the characteristic roached back line through the efforts of English breeders. The ancestral Malay undoubtedly had a form similar to that of the Shamo Games. The introduction of games of the Malay type was already undertaken in the first half of the last century, and must be judged as the prelude to the creation of new breeds of enormous vigor. Besides the principal color variety, gold-wheat-colored, eight additional colors are recognized, of which in recent years the tricolor spangled, white, pyle, black, and barred have occasionally been exhibited at shows. From Malay hens an egg production of up to 120 brownish eggs can be expected; nevertheless, the large-framed, robust body requires a great deal of concentrated feed. The eyes are yellowish-pearl-colored and the legs yellow; the comb is a strawberry comb.

The Shamo Game was specifically developed in Japan from games of Malay type. It is an excellent game with a robust body and widely spaced legs. The carriage is more upright and the neck is longer than in the Malay. In contrast to Malays with their strawberry comb, which is similar to a walnut half, in the Shamo a flat, three-rowed pea comb is required. In both breeds the wattles should be developed very little, in order to offer the opponent as few points of attack as possible when fighting. The Shamo hen produces two to three clutches of up to 20 eggs each annually; she hatches dependably and watches over her chicks for a very long time. Every strange creature that comes too close to her offspring is attacked.

For all game breeds the strongly developed musculature, particularly the breast and thigh areas, are characteristic, and herein — besides their significance as regenerators of the vigor of other chicken breeds — lies their special, commercially important traits. In Japan, various game breeds, from the largest to the smallest, are classified as "Shamo," while we only call the largest Japanese games by this name.

The Aseel, from the Indian region, also shows great regional variation, and in its homeland is given an additional regional designation. For our shows only Aseels of the "Rajah" strain are approved. They are very muscular and have broad, upright carried bodies on stable, barely medium-high legs. In this breed the fighting instinct is strongly pronounced even in the hens, so that swollen heads are an everyday occurrence in a fairly large holding. These games, just like noble horses and elephants, were kept for thousands of years in Indian royal palaces. A bird was seldom given away if it came from a successful bloodline, and the constant, thorough selection for fighting strength and vigor led to almost complete inbreeding in this breed. The hens often produce only two clutches a year and raise their own offspring in an exemplary fashion. It is a pure fighting chicken with cultural significance. The head is proportionally small, but the beak is nevertheless powerful and curved like that of a bird of prey; the comb and wattles are barely developed, the three-rowed pea comb only to a small degree in the cock. As is true of almost all game breeds, vigor, type, and form take precedence over color and markings. In the Aseel, bright-red, black-red, and pheasant-brown are the most commonly seen colors. Yellow leg color and eye color as light as possible (white) are standard requirements.

In England, specifically in Cornwall, the most massive game, the Indian Game, was bred from the Aseel. Because of their high body weight and their low, wide posture they are not suited for fighting, even though their heavy beak can take a frightful bite. Because of their heavy musculature and the rapid weight gain, the Indian Game (in the U.S.A. known as the "Cornish") are of great significance in commercial poultry breeding for breeding the broilers that are so familiar today. Indian Games are heavy, very wide-postured and enormously heavy-boned chickens with short, almost meager tail feathers and a robust head on a proportionally slender neck. As in all games, red ear-lobes are also required in Indian Games; the eyes should be as close to pearl-colored or light yellow as possible. In the particularly heavy birds the head points — particularly the three-rowed pea comb — which must be short, are also often relatively strongly developed. The pheasant-brown color with its uniquely beautiful feather with double-laced markings was first perfected in this breed, and was later reproduced in other breeds. In addition, we often find birds at shows in the red-white color variety (called "jubilee"). In practice, these two color varieties are crossed by breeders in Europe, and good show birds are obtained (in the U.S.A., a similar color pattern such as white-laced-red is better known). Approximately 60 to 80 brownish eggs can be expected from the hens of this breed. With broodiness, however, it can also be less. On account of the heavy weight, the hens are not particularly good sitters.

Games were also bred in Europe — even though later and at

Full ear-tufts and a beard are breed characteristics of the Orloff cock.

The Sussex rooster is especially attractive because of its strongly contrasting colors.

first of a different type. In these the Junglefowl type was predominant; games in Spain and Portugal, as well as in western and northwestern France, Belgium, and England are of a similar type. The most familiar are the Old English Games, which today are also bred for showing. Since the Romans already found gamefowl in England when they arrived, it can be assumed that these games of similar type originated from chickens that were carried along by seafaring traders. In contrast to the previously described games from the Far East, these games are only fast spur fighters, whose fights are often already decided in the first half minute.

After the appearance of the Malay in Europe, these were crossed with the indigenous games in Belgium, which resulted in the creation of the giant, heavy Belgian Game, of which the regional varieties Liege and Brugges Games are accepted for German shows. The Liege Game is somewhat slimmer and carries the body somewhat more upright than the Brugges Game; additionally, this variety should only have a single comb. Through crossbreeding of the two varieties in Belgium the distinction has become somewhat blurred; nevertheless, exhibition breeders have again produced single-combed flocks through strict selective breeding. The roosters' combs and wattles are usually dubbed before the onset of sexual maturity so that the pugnacious fellows cannot injure themselves too severely. Brugges Games should have flat pea combs; the eyes of both breeds should be as dark as possible and the legs blue-green. The elongated head is equipped with a powerful beak which is curved at the tip. A unique trait is the presence of spurs on the strong legs in some hens as well. The hens are credited with diligent laying with high egg weight. At shows, birds of the various common colors, such as birchen, brown-breasted, gold-necked, and silver-necked, as well as their blue variations or also solid blue or black, are exhibited. If the plumage is predominantly black, the legs are also correspondingly dark and the facial skin is purplish.

Of the European games, the Old English Games have achieved the widest distribution. Through the British colonies, in particular, these chickens became popular throughout the world. Consequently, independent varieties were developed in several countries, such as in Australia or in the cockpits of the United States. The Old English Game is one of Europe's oldest chicken breeds, and comes in over thirty color varieties in England. In the past they were looked upon purely and simply as the English chicken and were found there on many farms. Besides their robust constitution and satisfactory egg production, the excellent meat of these firm-fleshed birds was appreciated everywhere. The rooster's comb and wattles are dubbed shortly before sexual maturity. If only one rooster is kept, then this measure can be dispensed with. A dubbed rooster, however, always looks significantly truer to breed. Old English Games have a lively temperament, but on the other hand are very skeptical of strangers. If they are given enough attention, however, they soon become docile toward the keeper and take treats from his hand. They sit dependably and defend their chicks even against birds of prey. If eggs are collected regularly, the brooding drive can be held within limits. Up to 200 eggs a year can be laid by good hens, in exceptional cases even more.

This very muscular, firm-fleshed chicken only has a medium-length body, similar in form to a medium-size landfowl. Viewed from above, the body is heart-shaped, widest at the front, and especially striking because of the muscular wings. The tight-feathered, very glossy plumage allows the body's contours to show up well; the same is true of the muscular thighs, which are set well apart. The proportionally short, sharply tapered form with the relatively long neck, full neck fringe, and straight tail feathers with firm and well-curved sickle feathers of the tail area (in the male) endows this very mobile breed with a balanced appearance. The large number of color varieties is fascinating in its brilliant diversity; almost all possible color combinations are recognized. In recent years the color varieties golden-necked, silver-necked with orange back, birchen, and orange-breasted, as well as their blue variations and occasional pyles, were exhibited at shows in West Germany. The eye color of these color varieties is required to be fiery-red, except for those with predominantly black feathers. In the darker varieties, a dark eye color with a dark-red face and black-green to black-gray leg coloration are required. Otherwise, the named varieties should exhibit willow-green legs, which turn gray as a result of the loss of the yellow fat pigment in good laying hens. The small, erect single comb and the wattles consist of fine tissue. According to the standard description of their homeland, besides the leg colors already mentioned, white and yellow legs are also permitted in almost all color varieties.

About 1850, the Modern English Games were developed from the Old English Games through crosses with the Malay. The body shape of this breed, which is comparable in height to the Malay, with its angular shoulders and its flat back, is like a flattening iron. The short, tight-feathered plumage, the narrow-feathered, level tail, and the long, slender neck with the delicate, elongated, and flat head, as well as the slightly sloping carriage, are characteristic of this elegant, streamlined game. Nevertheless, interest in the Modern English Game has always been confined to a small circle of ardent sporting breeders. The color varieties shown in West Germany correspond to those of the Old English Game; depending on the color variety, the legs are willow-green, yellow, or black-olive.

Other breeds owe their existence to the combination of Malays with indigenous chickens. The Kraienköppe, whose roosters were at first also used for cock fighting, originated in this fashion in the West-German–Dutch border region between Bentheim and Enschede. As an exhibition breed, they first appeared in 1925 at German poultry shows. They have a robust, landfowl form with medium-high carriage, and are bred in silver-necked and golden-necked color varieties. The game-like appearance, which is especially expressed in the head points, is a characteristic of the breed and must be obvious. The colors are brilliant and light; therefore, in both color varieties a narrow silver-gray or gold edging of the hen's back feathers is allowable.

Above: White Wyandotte Bantams are particularly appealing because of their good proportions.

Below: The Barnevelder Bantam is a good laying breed.

Kraienköppen are very good layers. Under optimal conditions 200 to 220 eggs are laid by a hen in one laying season. An unlimited run, where a great deal of natural food is taken, is an advantage with this breed. The head, which has a small strawberry comb and small wattles, offers very few points of attack to the weather and should be bright red. The eyes should be as fire-red as possible; the legs should be yellow. The brooding drive is barely present in this breed.

A breed that is particularly striking because of the strong head points is the Orloff, which comes from Russia. In appearance they lie between Malays and robust, bearded barnyard chickens, and in their area of origin — the European region of the Soviet Union — they were also once used for cock fighting. The upright carriage, the broad head with the robust, curved beak, and the protruding eyebrows, as well as the thick beard and the full ear-tufts, give this breed a unique character. They were first imported to Germany in 1884. As a result of their full plumage and the small strawberry comb, they are relatively insensitive to cold weather.

Rhode Island Red Bantams are the most productive overall.

In West Germany the tricolor variety, whose markings on a mahogany-colored ground color resembles a coarse spangled color (feather tips black with white spots), is the most common. In recent years increasing numbers of mahogany-colored and white Orloffs have been exhibited at the large shows; in addition, black and barred varieties are also recognized.

In past years, two small game breeds from Japan were added to the poultry standard: the Tuzo and the Yamato. These are pure sporting breeds, which because of their small size and their quiet temperament are also satisfied with small runs. One can leave the incubation and rearing of the chicks to the dependable hens without concern. While the Tuzo of the same size is altogether slimmer and more elegant, the Yamato impresses because of the wide, upright, muscular body, and its very short plumage. The small, three-rowed pea comb sits close to the wide skull; older Yamatos produce a particularly odd impression because of their knotty, wrinkled faces. A characteristic of the breed is the extraordinarily fine scaling of the shanks.

The longtail fowl, which are closely related to games, are an unusual group. The Sumatra, from the Sunda Islands, is an ancient breed. This breed still embodies the game type to the greatest degree; it stands slightly over medium height and should exhibit extremely luxuriant, full, yet tight-feathered plumage. The stiff, long sickles should only sink downwards in a gentle arc in their second half. The deep-black plumage with the beetle-green sheen unique to this breed has been carried over to many other breeds through crosses. The Sumatra has a slender, pheasant-like form. The head points are predominantly mulberry colored, even after the onset of sexual maturity; occasionally the pea comb and sometimes the wattles are red at full maturity. An egg production of about 100 to 120 can be expected from this breed, which has great ornamental value and cultural significance. The hens sit and raise the chicks quite well. The eyes should be dark reddish-brown and the legs should be olive-black; the sole of the foot must be yellow.

From the Caribbean — from Cuba — comes the Cubalaya Game, which was refined in the United States and which was introduced at poultry shows in West Germany only in recent years. Like many games, this one also has the characteristic pea comb. A peculiarity of this breed is the dragging tail, which is carried downward and spread out, and ideally should resemble a lobster tail. The somewhat compact body has a sloping carriage and is of medium height. This breed is available in a cinnamon color and should be classified as a sporting chicken.

The Yokohama from Japan was already known at the start of organized poultry breeding. Only if one compares illustrations in old poultry books with our present-day Yokohama, however, can one truly recognize the how these noble longtail fowl were perfected by European breeders. Besides the elegant, upright form, the red-saddled color variety with the fine pearl markings on the red feather areas of the otherwise white chicken is a masterpiece of the art of breeding. This breed also occurs in the white color variety. The small walnut comb, the ear-lobes, and the barely visible wattles at the throat are red, and together with the required red eyes and the yellow legs and beak form a beautiful contrast. On grass especially this breed is a pleasure to behold. In addition to the great ornamental value, one can expect about 80 to 100 eggs a year from the hens if they are given proper care.

The ultimate longtail fowl is the Phoenix. In West Germany as in Japan — its homeland — it occurs in two breeding lines: the normally molting (replacing the plumage) Shokuku and the Onagadori, in which the sickle feathers and the saddle feathers of the rooster continue to grow under special housing conditions and become about one meter longer each year. In Japan these breeds are preserved and supported at special breeding stations as a national cultural monument. Private breeders are supported with monetary awards; the most beautiful roosters, with tail lengths of up to 17 meters, can be seen in the breeding stations for a fee. Special cages, which protect the tail feathers from soiling and damage, have been developed for housing these roosters. Caring for these birds is very expensive and can appeal only to a few. The hens molt yearly and can be kept in normal chicken stalls. The breeding cock's tail usually reaches a length of only one meter, since the feather tips wear away or break off during normal movements. The hens of this ornamental breed produce at most two or three clutches with a total of about 60 eggs a year. Color varieties and head points are not judged as strictly as with the Phoenix-Shokuku described in the following, in which, besides the allowable color varieties gold-necked, silver-necked, orange-necked, and white, one also finds brown and black-red ones.

The Phoenix-Shokuku, once miniaturizing the breed attracted many new interested parties, has been greatly improved in vitality, form, and size in the last decade. This has also had a positive effect on egg production, although profitability cannot be expected of an ornamental breed. The Phoenix's head points are required to be relatively small and delicate, the ear-lobes should be white, and the comb should be fine and provided with short teeth (on the upper third of the comb blade).

The hens have a slim landfowl form and carry the generously developed tail feathers only slightly above horizontal. The plumage coloration should always be brilliant. This breed's tail feathers are especially abundant; it is not so much a question of the width of the individual feathers, but instead of the highest possible number of strong, flexible, long feathers. This decisive characteristic of the breed should be of primary importance in breeding as well as in judging at poultry shows.

Breeds of Asiatic Type

The greatest revitalization of poultry breeding began in 1849 with the importation of the heavy Asiatic chicken breeds, primarily the Cochin. In subsequent decades one spoke of a real "Cochin craze," which regrettably caused a number of established native breeds to fall into oblivion. Of course the first "Cochin Chinas," as they were called in those days, had little in common in appearance with the

In the Silkie Bantam, the feathers are very obviously shredded. The plumage is reminiscent of the coats of mammals.

A striped Wyandotte Bantam cock exhibits the most prominent characteristic of the breed.

Cochins exhibited at shows today. The decisive, commercially important trait of this new breed, besides the size and the high weight, was the ability to lay in the winter. Through crosses with lighter, indigenous breeds, completely new ones with the best production traits were developed. The Cochin, as known to us today, is a large, massive chicken with profuse plumage and a full and deep body on strong, heavily feathered legs. The head has a proportionally delicate single comb, the ear-lobes are red, and the wattles should be very well developed. The eyes must be orange-red, and the legs should be yellow; if the plumage coloration is predominantly black, a willow-green leg coloration is also permitted. No great expectations may be placed on the egg production of the often broody hens; good hens lay 120 to 140 yellow-brown eggs. On the other hand, this breed is very meaty.

The Brahma was developed from the Cochin and the Malay in North America in about the middle of the eighteenth century. From the Malay they received the upright posture and the delicate head points; the Cochin provided the full body, the abundant plumage, and the feathered legs. This breed is currently by far the largest in poultry shows. The roosters can weigh over five kilograms, the hens over four and a half. About the same egg production can be expected as with the Cochin; the principal advantage of this breed as well is the amount of meat. The coop, its furnishings, and the run must correspond to the size of these birds. Because of their docile nature, each bird does not have to have more area than was previously given, but because of the abundant feathering of the feet the greatest attention must be paid to the care of the run. This breed's pea comb and wattles are small; between the wattles there is a well-developed fold, which is characteristic of the breed. The heavy-boned legs must be a rich yellow; the eyes should be orange-red. At present, this breed comes in five color varieties.

The Langshan, which was first imported around 1872 to England from northern China, and the Cochin, which comes from southern China, presumably have common ancestors: the similar, upright station and the more or less pronounced leg feathering point to this conclusion. In Germany, through crossing with the Minorca, among others, a unique smooth-legged breeding line was developed, which soon attracted many fanciers and in Germany almost completely supplanted the English "Croad Langshan" breeding line. Only in recent years has the Croad Langshan again become somewhat more popular. The German Langshans are bred in the color varieties black, blue-laced, brown-red, and white, the Croad Langshans in black and white. The plumage of the black variety has a beautiful, beetle-green sheen. The legs of the blacks and blues are black to blue-gray, the eyes are black-brown; in the whites, the leg color is slate-blue to bluish-white with orange-red eyes. The Croad Langshan, in particular, is known as a layer of large, brownish eggs; according to a number of breeders, very good hens lay 180 or more eggs a year. Because of the large-framed skeleton, this breed should also be provided with ample amounts of minerals while young. Highly recommended is supplemental feeding of bone scraps or the remains from the sawing of bones from butcher shops.

In the second half of the last century, a breed with significant commercial traits was bred by an especially talented English breeder. This breed, the Orpington (named after its locality of origin), was developed from the Cochin, Langshan, Minorca, and several other breeds, and is an all-around utility chicken. The fast growth is particularly praiseworthy. In addition to excellent egg production, the meat is also outstanding. The once-welcome brooding drive is not as highly valued by breeders today, since egg production suffers. The cubic form, typical of the breed today, was developed through the continued efforts of breeders. To this now belongs a low station; nevertheless, there should be sufficient clearance between the plumage and the ground. Because of their tendency to become fat quickly if given concentrated feed, Orpingtons should always be fed sparingly. Laying hens are best given only laying meal in unlimited amounts, and for the grain ration should be given a small amount of oats (20 grams a day for each). Fat hens lay poorly and produce hardly any fertile eggs. Orpingtons have a small single comb, red wattles, and orange-red eyes. The legs are flesh-colored to white, or, with dark plumage, gray-black (in this case the eyes are dark as well). The skin of these chickens is also white. Ten color varieties are recognized for poultry shows.

After the First World War, the Australorp was bred in Australia from the Orpington for high laying production. After this breed was introduced to poultry farms in the 1950s as a commercial breed, it was also simultaneously improved as a show chicken, although only the black variety was recognized until a few years ago. Today, Australorps can be found at almost every poultry show, since they provide a good roast in addition to the high egg yield and they always look dapper in their black plumage with its green sheen. A few years ago this breed was also approved in white. As is true of all medium-weight breeds, Australorps do not require a large run, and the fencing need only be one and a half meters high. Nev-

The distinctive feathering on the neck of the Greylag Goose (above) persists in the domesticated breeds that stem from it, just as those of the Swan Goose (below) preserve the overall markings.

Facing page: Domesticated waterfowl quickly become partial to being fed bread by hand.

ertheless, green food is always taken gratefully and serves to maintain health.

The Plymouth Rocks, which were developed around 1860, are an American breed. They were robust, hardy production chickens and quickly became popular in Europe as well. The initial barred color variety was constantly improved in color and markings; this was accomplished at the cost of production in this color variety. Although they continue to be the most popular variety because of their perfect barring, the buff, white, and black Plymouth Rocks mature earlier and are somewhat more productive. In addition, this breed, with its small single comb, red eyes, and yellow legs, is also recognized in the partridge color of the Asiatic breeds; on the red-brown ground color, the hen's body feathers, in particular, have several narrow, black bands. This color variety is regrettably rare. The large hens lay 180 yellowish to brownish eggs in the most productive color varieties; occasionally, higher yields have also been achieved. The meat yield is also satisfactory. The birds have a very quiet disposition, an elongated form on a proportionally wide carriage, and only a short tail area.

A newer commercially important chicken breed with somewhat coarser striping is the Amrock. As is easily recognized from the name, these chickens also come from the United States. In this breed a high egg production and a respectable meat yield is firmly established as a result of decades of selective breeding. This medium-weight chicken with yellow legs, small single comb, red ear-lobes, and orange-red eyes presents a cheerful sight on a green lawn, thanks to its gray-white barring on the black ground color. With an optimal diet, hens of this commercial breed can lay 220 eggs, first-class hens up to 250 eggs in a laying year. The eggs are a yellow-brown color and medium size; broodiness is barely present. The sexes of the chicks of this breed can be told apart on the first day by the down plumage; the leg color of the one-day-old rooster is lighter and the light head patch about a third larger than in the one-day-old hen. This breed also has a quiet disposition and requires only a low fence. In all barred and striped breeds the roosters are lighter in coloration than the hens, because they possess two genes for the light diagonal striping on the feathers. Amrocks also mature early: the hens begin laying at an age of five months.

The Sussex, whose past commercial importance lay in its all-around economic value, comes from southern England. In addition to the excellent fattening ability, the good egg production is also praiseworthy. Toward the end of the last century, the broodiness of this breed with good winter laying was valued, particularly by smaller poultry breeders. Because of the use of incubators and artificial rearing, the brooding drive became superfluous, and through selective breeding this characteristic has been largely suppressed. In this way the annual egg production could be increased. Hens of the popular light color variety produce an average of 200 yellow-brown eggs a year. Characteristic of this breed, which is also bred in the color varieties red Columbian, buff Columbian, spangled, and brown, is the heavy box-shape. The upper and lower body lines run parallel and almost horizontally. The station is medium high, the legs are flesh-colored-white, and the head has a barely medium-size single comb. The ear-lobes must be red, the eyes orange-red. The Sussex is also relatively widely distributed internationally. The Columbian marking consists of black stripes on the feather shafts of the hackle and predominantly black tail plumage. The tail area is only of medium length in this breed.

In the second half of the last century, particularly in the vicinity of economically important cities and correspondingly wealthy populations, other chicken breeds were produced through hybridization in order to supply the demand of the well-to-do for tender, young poultry. The Sundheim Fowl originated in this way in the district of Kehl in Baden. This former meat chicken developed further into a recognized breed, while the majority of regional meat-chicken breeds died out because of the difficult times of the World Wars and the subsequent years, along with the advent of commercial poultry breeding with its specialized hybrid breeds. The Sundheim received its coloration (Columbian marking on a white ground color) and its legs, which are equipped with small feathers on the outside, from crossing with the Brahma. Today these are fast-growing, so-called dual-purpose chickens (used for both egg and meat production), which are somewhat lighter and not as high as the previously described Sussex. Sundheims have a small single comb, red ear-lobes, and orange-red eyes; the legs must be flesh-colored. The medium-length body must be broad and deep, so that enough meat can be put on. Since the brooding drive has been curbed through the efforts of breeders in this breed as well, from well-bred hens 180 of the beautiful brown eggs can be expected. As is true of almost all dual-purpose chicken breeds, a small run with a fence only one and a half meters high is sufficient for the Sundheim. To prevent soiling the predominantly white plumage, a grass run is advisable.

The Lachshuhn was bred as a dual-purpose chicken in Germany from the old French meat chicken breed, the Faverolles. The Faverol-

Above: The majority of domestic geese have Greylag Geese as their ancestors.

Below: Diepholz geese were bred to be grazing animals especially.

Facing page: A young goose with flight feathers just growing in.

les, as a result of crosses, apparently contains the blood of various good meat breeds, such as the Brahma (feathered legs), the Houdan (beard formation and a fifth toe), and the Dorking (five toes). The German breeding line exhibits, in addition to good fattening ability, a satisfactory laying production, and it has a correspondingly wide popularity. Particularly striking in this breed, besides the low, trapezoidal body, is the salmon coloration that gives it its name. In this color variety the rooster and the hen are differently colored to an unusual degree. While the rooster possesses predominantly black plumage with straw-colored neck and saddle feathers and brownish-red, brass-colored laced back and shoulder plumage, the hen does not have any black at all in her plumage. She is creamy-white on the underside; neck, back, and wing feathers are salmon-red with whitish lacing and feather shafts. The contrast between the hen's whitish beard and the rooster's black one is particularly interesting. In addition, a white color variety is also permitted. The shanks are white, the eyes must be orange-red. The single upright comb is small, and the vestigial wattles are covered by the full beard. These docile chickens look particularly attractive on a mowed lawn.

The Mechelner is a Belgian meat chicken breed. This breed, primarily bred in the barred color variety, has become extremely rare. These meat chickens once had a great reputation as "Brussels Poulards." The large, broad, deep body is carried horizontally and stands on proportionally short legs which are feathered on the outside. This breed also exhibits a small single upright comb and red ear-lobes, as well as orange-red eyes. This breed's egg production is secondary to the breeding goal of meat-chicken traits. Besides the coarsely barred, the white color variety is also recognized at poultry shows. The flesh-colored legs may have a gray tinge in the hens.

The blue-barred, today called the Niederrheiner, which were bred from the northern Dutch blues, have been able to attract a fairly large circle of breeders. This breed was expanded in recent years by the blue color variety; in addition, the color varieties yellow-barred, sex-linked, and birchen are recognized. This breed embodies a productive dual-purpose type with only a medium-long but deep body and a rounded, rising back line. Medium height is required; the head points with the single upright comb are well developed. The eyes are orange-red, and the legs, depending on the color variety, are flesh-colored with a varying degree of a dark suffusion. Apart from a good meat yield, this breed provides proportionally high egg production; the shells should be yellow-brown.

Wyandottes were bred in the United States toward the end of the last century. Silver Wyandottes were the first color variety recognized. Even today these enjoy comparatively wide popularity; presumably the dapper markings (white ground color with black lacing) has something to do with this. In no time at all the white color variety had a worldwide distribution as a commercial chicken. Wyandottes have a rose comb (which is insensitive to weather), yellow legs, and red eyes. The large body is well-rounded on all sides and provides sufficient room for the organs essential for production. This breed is recognized in almost all poultry colors, and in this respect offers something for every taste. As is so often the case, several of the former commercially important color varieties were bred one-sidedly for "exhibition beauty," and thus lost some of their productivity. In this connection, one should inquire exhaustively about the desired traits before purchasing a breed from a breeder. Birds whose bodies are too short and cylindrical usually have difficulty during reproduction and can hardly provide the desired egg production either. Depending on the color variety, the color of the eggshell varies between cream-yellow and brown. The egg production is between 140 and 220 eggs a year.

One breed has been bred for special commercial traits in Germany since 1907: the German Reichshuhn. In order to provide environmental conditions as few points of attack as possible, a small rose comb, which lies close to the skull, was developed. The form is elongated and brick-shaped, the carriage flat, and the tail is relatively short. The eyes of these chickens are orange-red and the legs are flesh-colored, since at one time white-skinned poultry were preferred. This relatively productive breed is recognized in the color varieties white, light, buff Columbian (black tail feathers and neck markings), striped, red, black, and silver. One can expect 180 to 200 eggs per hen per year. Because of the limited flying ability, the fencing need only be one and a half meters high.

In the Netherlands, for commercial reasons and to satisfy the demand for the popular dark-brown eggs, chicken varieties with very good laying characteristics and a good meat yield were bred by crossing various breeds of Asiatic origin. Two dual-purpose breeds which are very popular today were developed from this racial mixture. The first breed is the Barnevelder, whose most popular color variety is the double laced. Here each feather has two brown markings on a black background, which are enclosed by the double black lanceolate edging. In the rooster this marking must at least be present in concealed form;

Above: The Sebastopol is a dapper domestic goose. **Below:** The Chinese goose is one of the domestic goose breeds descended from the Swan Goose.

Facing page: Precocious goslings, though still covered with down, require little care from their owner.

if its feathers are viewed from below, they must look like the hen's feathers. This trait must be considered in roosters for breeding in particular. Otherwise, we find this marking only in the pheasant-brown Indian Game. The Barnevelder is of medium weight and has a roomy build; the form is somewhat compressed, and the line of the back is concave and rises without angles. In addition, the broad body allows optimal meat production. A single upright comb sits on the head, the eyes must be orange-red, and the legs should be yellow. A dark suffusion is usually present on the legs, particularly in double-laced hens. Barnevelders are also recognized in black, white, and dark-brown. A shiny, dark-brown egg weighing at least 60 grams is a breeding goal. For this reason, care must be taken that the brood cock in particular is descended from a hen that exhibits this characteristic of the breed. Because of the high egg weight, the number of eggs laid in a year is slightly lower; good hens, however, lay about 180 eggs a year.

The Welsumers have a similar origin. They differ from the previously described breed in form and coloration. Their form is elongated-cylindrical. This breed has so far only been known in the rust-partridge color variety; however, orange-colored birds have been exhibited in recent years. The rust-partridge color represents a lightened partridge-color with a higher percentage of brown; in the rooster the outer black breast feathers are provided with a brown center portion. Head points and leg color are the same as the Barnevelder. The principal characteristic of this hardy, dual-purpose breed is the heavy, blunt, dark-brown egg; eggs of over 70 grams are no rarity. One must make sure that only those eggs that meet the breeding goal are incubated. Eggs that are too heavy are not suitable for incubation; instead, eggs with a weight between 60 and 70 grams should be sought out for incubating. Because of the high egg weight, Welsumer hens usually only lay about 160 eggs a year; with lower egg weights a greater number could occur. With this breed, as with many robust dual-purpose breeds, it is important that the birds are not overfed. It is also advisable to offer laying mash in measured portions and to give a short measure of grain in the evening ration in particular. According to experienced breeders, in this way better laying production and breeding results can be expected.

The United States has produced important commercial breeds which have attained worldwide recognition and are the foundation for many hybrid breeds of the modern poultry industry. Besides the Leghorn, which was already described among the light laying breeds, these include the Rhode Island Red, which originated at the beginning of the second half of the previous century — primarily in Rhode Island — by crossing red Malay Games with the available indigenous chicken varieties. This breed was imported into Germany right at the start of the twentieth century, and, because of its dual-purpose nature (in addition to the high production of brown-shelled eggs, it also has a respectable meat production), it gained acceptance in commercially-oriented poultry breeding. This breed is characterized by high winter laying production; through breeding for ever more uniform and dark plumage coloration with intense sheen, the start of laying was somewhat delayed, however. Young Rhode Island Reds today begin laying eggs at an age of about six months. The breed is very popular and has a large circle of breeders. When purchasing Rhode Island Reds, because of the differences in production, stocks controlled in this respect should be given preference. When putting together a breeding flock, balance — particularly with respect to coloration — must be given consideration; birds with a dark and glossy color should be mated with birds of a lighter shade with healthy, broad feathers, since when mating two dark, glossy birds, many chicks will be predominantly unfeathered or will have hairy or loose feathers. Today, Rhode Island Reds can only be found in the dark-red-and-white color; the comb is a single upright comb. Rose-combed Rhode Island Reds are, however, also permitted. The eyes should be orange-red and the legs must be yellow. In a good flock, production of about 200 eggs with brownish shells can be expected from each hen. The birds fly poorly, so that the fence need only be about 120 to 150 centimeters high.

In the United States, a perfected production chicken — perfected in laying as well as in meat production — was developed from the Rhode Island Red. This breed is the New Hampshire, which also reached Germany after the Second World War and soon found a large circle of breeders. It is a fast-growing and early maturing chicken with robust health. Originally these birds had only brown plumage; this was refined through crossing and selection in the gold-brown direction, in order to make this breed interesting for show breeders as well. By the 1950s the white color variety was simultaneously being bred in Germany and the United States. It has, however, never attained the popularity of the gold-brown variety. Besides the compact form with the slightly concave and rising back line, the New Hampshire is characterized by a single upright comb, which is somewhat larger than with its ancestor, the Rhode Island Red. As with the Rhode Island Red, the legs are yellow and the eyes are orange-red. In the gold-brown variety,

Above: Almost all domestic duck breeds derive from the Mallard.

Below: The Pekin duck, with its attractive all-white plumage, is a popular utility duck.

Above: Campbell ducks are very good egg producers.

Below: In ducks, the "crest" looks like a ball of feathers.

the black tail plumage and the light black drop-shaped markings of the neck feathers in the hens are especially attractive. Despite the breeding for beauty, production of 200 or more large, brownish eggs a year can be expected from New Hampshire hens. The young hens often begin laying at an age of four-and-a-half to five months. New Hampshires may be given only a small amount of grain; 30 grams per bird per day is quite sufficient. Laying mash may be offered freely. If given diets rich in grain, the chickens fatten very quickly and their production suffers.

In the 1950s, a new breed, the Dresdener, was bred from the Wyandotte, Rhode Island Red, and New Hampshire in East Germany. It combines the productivity of the Rhode Island Red and New Hampshire with the Wyandotte's rose comb, which is insensitive to cold, and it also has a very attractive, balanced shape. In the meantime, Dresdeners have been recognized in four color varieties: besides the gold-brown and white birds bred initially, there are also black and, for several years, also rust-partridge ones. The high laying production has secured this breed a permanent circle of breeders.

A new chicken breed of dual-purpose type was developed in the 1970s in the area of Bielefeld, West Germany. This very productive chicken received the name Bielefelder Sex-link, for the one-day-old chicks can already be clearly separated into males and females on the basis of their differing down coloration. This large chicken has an elongated frame with a long, straight back, and scarcely shows thighs at all. One can almost designate the robust body as thickset. A simple upright comb adorns the head; the eyes must be orange-red and the legs should be yellow. The Sex-link's color can best be described as follows: on a rust-partridge-colored ground color, a light-gray barred marking is distributed over the entire body. This breed is very fast growing and must not be fed too heavily with grain; the same recommendation applies here as with the Orpington or the New Hampshire. With an optimal diet, up to 220 heavy brown eggs can be expected from each hen. On the basis of production and the certainly quite attractive exterior, rather wide popularity can be predicted for this breed.

Bantams

Miniaturized Breeds

Two large groups of bantams exist: on the one hand the miniaturized breeds, on the other hand the true bantams (original bantams). With the former, in principle it is a matter of nothing more than the large breed in a reduced size. To be sure, there are more or less large differences in detail; nevertheless, these are usually evident only to judges at poultry shows. For this reason, in the description of bantam breeds the miniaturized breeds will be omitted; information about these breeds is given under the respective large breed. Of course, not every large breed has a corresponding miniaturized breed. The following miniaturized breeds are known: Malay, Indian Game, Belgian Game, Old English Game, Modern English Game, Orloff, Kraienköppe, Yokohama, Phoenix, Brahma, Langshan, Orpington, Australorps, Plymouth Rock, Amrock, Sussex, Lachshuhn, Sundheim, Niederrheiner, Wyandotte, German Reichshuhn, Barnevelder, Welsumer, Rhode Island Red, German Barred, New Hampshire, Dresdener, Araucana, Naked Neck, Sulmtaler, Altsteirer, Vorwerk Fowl, Castilian, Andalusian, Minorca, Italian, Leghorn, Polish, Houdan, Brabanter, La Fleche, Augsburg, Rheinlander, Mountain Crower, Braekel, Ostfrisian Moven, Lakenvelder, Hamburg, Bearded Thuringer, and Rumpless Fowl. A breed not represented in the large type is the Ruhlaer Bantam. It belongs to the "true bantams" discussed in the following. The weight of bantams varies between 800 and about 1000 grams. The egg weight lies in the range from 30 to 40 grams.

True Bantams

Completely separate from the miniaturized breeds (that is, not represented in the large breeds at all) are the original bantams, or true bantams.

Silkie Bantams are the first of these breeds. As can be proven, they have been known for 700 years. According to Marco Polo's travel reports, the Chinese kept the Silkie Bantam as a domestic animal. It reached Europe around 1780, and then Germany in the following century. Since it cannot fly, it is virtually ideal as a chicken for enclosures with low fencing. Its tameness alone makes it popular. Many breeders also value its brooding drive and its brooding reliability. Its principal characteristic is the silky, shredded and hair-like plumage from which it takes its name. With respect to the form, it resembles a rounded cube. The legs and the outer toes are lightly feathered ("stockinged"). A peculiarity is the fairly long fifth toe directly over the hind toe. Under the plumage, which occurs in the colors white, black, blue, black-red, and silver-gray, it has a dark skin. The flesh is also much darker than in other domestic chicken breeds. In flavor it inclines in the direction of gamebirds. A well-developed crest sits on a cranial protuberance. On the head it carries ear-tufts and a beard, but it can also be totally beardless. A peculiarity is the mulberry comb. It is a diagonally lying, mulberry-shaped comb with a bluish-red coloration. The rooster weighs about 1,000 grams, the hen

200 grams less. The eggs weigh 35 grams. The eggs are not well suited for artificial incubation, but do very much better when incubated naturally.

The origin of the Bantam Frizzle lies in obscurity. Historically, it turns up in many places. The feathers are "shaggy" (that is to say, crinkled or crumpled), are turned upward, and have a supple structure, so that they resemble those of the Silkie Bantam to some degree. According to breeders, to strengthen the beautiful shaggy structure, one should cross in a smooth-feathered rooster from time to time.

Although the Cochin also exists as a large breed, the bantams have nothing in common with the large fowl. They come from China and were kept there as a domestic chicken for a long time. Only in this century did this breed experience a considerable increase in popularity in Germany. Because of its tame and confiding disposition, the Cochin Bantam, like the Silkie Bantam and the Frizzle, is suited to an enclosure with a low fence, and it makes an especially attractive impression on a grass run. One cannot expect productivity from this breed. Its overall appearance gives the impression of a round ball of feathers;

Below: "Dainty" seems to be the word for the Call duck, which weighs only about one kilogram.

Facing page: Muscovies are inclined to forage in the tall grass for the insects found there.

that is, the outer boundary of the body should fit inside a ball. The feather structure is supple and at the same time strongly pronounced. The quite short legs are equipped with long feathers on the outside, as are the outer and middle toes. A small single comb adorns the head. The Cochin Bantam occurs in a broad spectrum of color varieties. Buff, white, black, mottled, black blue, pearl-gray, barred, buff-barred, partridge, brown-laced, birchen, dark, silver-laced, light, buff-Columbian, and wheaten are the color varieties. At 850 grams for the rooster and 750 grams for the hen, this bantam is quite light. The egg, which weighs 30 grams, is a brown color.

An especially curious contemporary is the little Chabo (called "Japs" in the U.S., short for "Japanese"). The rooster weighs 600 grams, while the hen is about 100 grams lighter. The breed's age is estimated at over 1,000 years; it originated in China and Japan. Conspicuous in these small fellows is the short-leggedness, which presents problems during mating. Short-leggedness in purebred form is also a lethal factor. This means that the embryo in the egg develops only to a certain stage and then dies. For that reason, one must also work with long-legged birds when breeding. The overall impression is that the legs are so short that the body almost brushes against the ground. The wing tips touch the ground. Both sexes carry the tail at approximately a right angle, whereby in the rooster the main sickles extend past the head by a third. The comb is relatively large and sometimes touches the tail in the cock. In the hen the comb can be tipped over. The following color varieties are recognized in Germany: white, black, blue, Siro (white with black tail), buff with black tail, Butschi (mottled), barred, black-golden-necked, black-silver-necked, partridge, dark, silver-wheaten, duck-wing cock, gold-wheaten, and spangled.

In all likelihood, the Javas come from the island of Java. Already in the second half of the nineteenth century these original bantams were imported into Germany in the black color variety. The additional color varieties white, barred, mottled, light, buff, blue, tricolor spangled, buff-Columbian, and golden-necked were bred in England and Germany as well as the U.S. This bold and temperamental chicken exhibits flowing and round body shapes. The rooster has a fully sickled tail, whereby the sickles are bent into the shape of a crescent. The drooping wings are characteristic. As head points, the large, white, round earlobes as well as the relatively large finely beaded rose comb are conspicuous. If the rooster and hen are too large, then the birds lose their elegance and style.

In the nineteenth century, the English horse and cattle breeder Sir John Sebright bred the bantams that are named after him. Flowing and harmonious lines are required here as well. Angles and edges are disruptive in this graceful chicken. The rooster has no pronounced ornamental plumage, which is why one speaks of "hen-featheredness" in the cock. The tendency to fly and the temperament can often become annoying in the Sebright. The laying production is about 100 eggs a year, and the egg weighs 30 grams. The carriage of the wings and the head points greatly resemble those of the Java, although the ear-lobes are significantly smaller. As color varieties, gold and silver are allowed. Each feather carries a black, gleaming green lacing.

The homelands of the ancestors of the Bearded Antwerps were Belgium and the Netherlands. The final breeding took place in Belgium. The rooster and hen let their wings droop. The rooster carries the tail very high, almost vertically; a somewhat lower carriage is exhibited in the hen. The face is adorned by full ear-tufts and a beard that covers the wattles. In the nape region the neck plumage curves into a mane. Above all, the form of the full, round, protruding breast looks impressive. A wedge-shaped rose comb with beading lends this 600 to 700 gram chicken a special radiance. The recognized color varieties are black, white, pearl-gray, barred, quail, blue quail, mottled, spangled, buff-Columbian, blue-laced, red-saddled, and blue porcelain.

The Bearded Watermaals are very similar to the Antwerps. They were bred at the start of this century in Watermaal (Belgium). In contrast to the Antwerps, they have a rearward pointing tuft on the head, and the rose comb has a three-part comb point. The 700 gram rooster and the hen, approximately 100 grams lighter, are recognized in numerous color varieties.

The Roman agronomist Columella wrote about feather-footed (booted) bantams in 60 A. D. This is apparently not a comment on the D'Uccles breed, as these can be shown to have originated only in the eighteenth and nineteenth centuries in Europe. As the name states, their legs are heavily feathered. For this reason, these bantams require special care. The run should consist of a short lawn or sand. The chicks require more room than with other breeds, since the development of the foot plumage suffers when they are overcrowded. They are similar to the Bearded Antwerps with respect to weight as well as the tail and wing carriage. The head is not, however, adorned with a rose comb, but with a small single comb with uniform points. The face is featherless or covered by full ear-tufts and a beard. In temperament they are — like the Antwerps — bold but confiding.

The Wild Turkey, the ancestral form of our domestic turkey, lives in North America.

This chicken is represented in numerous color varieties. Recognized are mille fleur, blue porcelain, pearl-gray with white spots, pearl-gray, mottled, black, buff, buff with white spots, white, barred, light, gold-necked, silver-necked, birchen, and lemon porcelain.

From the region of Liege come the Basetten, which have been known since the 1930s. These are robust, barely medium-high bantams of landfowl form with upright carriage. In temperament, Basetten are very lively but also docile. With 160 eggs a year — whereby they have also proved to be good winter layers — they are outstanding laying chickens in bantam form. Their vitality also results in good hatching and rearing results. The broad body with the well-tucked-up tail characterizes this bantam. An upright single comb, which folds back in the rear in the hen, adorns the head with the white ear-lobes. With respect to color varieties, only quail and silver quail are recognized. The rooster brings 900 grams, the hen 800 grams onto the scale.

As early as the eighteenth century, the Hollands existed in their homeland. This old bantam was not, however, recognized as a distinct breed in northwestern Europe until 1906. The confidingness, ease of care, and vitality led to its popularity, particularly with breeders with little space. The weight of the cock and hen varies between 500 and 550

Above: Blue turkeys are quite rare, though their coloration is particularly pleasing.

Below: The remarkable display of the turkey cock results from each feather being controlled by a set of small muscles.

grams and 400 and 450 grams, respectively. The carriage of this well-balanced chicken with its full breast and well-tucked-up tail produces a plucky effect. The downward-pointing wings are a characteristic of the breed. The five-pointed single comb and the white, almond-shaped ear-lobes characterize the attractive head. A broad palette of color varieties exists: gold-necked, silver-necked, orange-necked, blue gold-necked, red-saddled, black, white, blue, barred, salmon-wheaten, and sex linked.

The elongated, pheasantlike German Bantam was bred in Germany from indigenous bantams and the Phoenix, and was exhibited for the first time in 1917. It carries the cylindrical, slim, long body close to horizontal, with closely held wings. Its body is twice as long as high. The 700 gram rooster has a slightly elevated tail with full, broad sickles. The hen, which weighs about 100 grams less, has a somewhat narrower tail which is slightly drawn in. The head is adorned by a small single comb, which resembles that of the Red Junglefowl, and heart-shaped ear-lobes. Its robustness and vitality have allowed this chicken, which occurs in many color varieties, to become very popular. The color varieties are black-red, gold-necked, orange-necked, silver-necked, blue gold-necked, blue silver-necked, red-saddled, white, black, spangled, light, buff-Columbian, and pearl-gray.

Geese

In geese we distinguish between heavy and light breeds. Before one decides on a type of breed, one must realize that while the heavy breeds bring more weight to the scales they also require correspondingly more feed. Birds with a weight of from five to six kilograms are the most profitable in the cost-to-yield calculation.

The homeland of the Embden goose is East Friesland. It has been bred for high weight production. Weights of 11 to 12 kilograms for the gander and 10 to 11 kilograms for the goose are considered ideal. On account of its mass, it made a considerable contribution to the popularity of goose meat in Poland and Russia. From the time the breed was developed until it attained its present-day form, it has experienced numerous crossings. This goose often produces 70 or more eggs a year. Unfortunately, the instinct to brood and rear the goslings has suffered because of this high egg production, so that eggs have to be incubated artificially or with the aid of turkey or chicken broody hens. The overall impression is of a large, white, heavy goose with a swanlike neck, a massive, long body, and a dual-lobed stern (hanging belly folds) which must be tucked up in the rear and may not touch the ground. The head is slim and elongated, the forehead flattened.

The Pomeranian goose, depending on the sex, reaches seven to eight kilograms. The brooding drive is relatively well developed in this breed, and it lays and hatches one or two clutches of ten to twelve eggs itself. The homeland of this goose, which occurs in the colors white, gray, and pied — as the name indicates — is Pomerania. In contrast to the Embden goose, this breed has experienced relatively few crossings. Especially prized is the meat on the thighs and the breast. In shape this is a large, heavy goose with a rounded, oval body and a deep, broad breast, as well as pronounced shoulders and a single-lobed stern. It carries its body horizontally. The head exhibits a clear forehead. In the gray-pied the head and half of the neck, the shoulders merging slightly into the wings, the lower back, the thigh plumage, and the tail feathers are dark gray. The tail feathers also have white edging.

The Toulouse goose comes from southwestern France, where it has occurred since the fourteenth century. It was perfected as a breed, however, in England. Although the weight is given as 9 to 10 kilograms for the gander and 8 to 9 kilograms according to the breed standard, with a suitable diet a weight of 12 to 15 kilograms can be attained. The greater the weight a goose carries, however, the poorer are fertility and laying production. About 40 eggs, which weigh about 200 grams, are normal for this breed. Although the brooding drive is still present in these geese, they do not prove themselves to be sufficiently dependable. The gray Toulouse is particularly well suited to supplying feathers. The overall impression is of a massive goose. The robust neck carries a broad, short head with a knob on the forehead; on the head itself can be seen a well developed dewlap (the feathered fold of skin on the throat). The dual-lobed stern, closed in the rear, reaches to the ground.

The first of the light breeds in Germany is represented by the Diepholz goose, which was not bred for high weight but for mobility and which brought great advantages with respect to pasturing. At the same time a high value was placed on maintaining the broodiness and on dependable care of the young goslings. This breed usually begins laying in the fall. If the eggs are removed, they lay second and third clutches. The hatched goslings are quite robust and hardy. The sometimes rough Saxon region of Diepholz, where they originated, contributed to their robustness. Despite their classification as a light breed, this goose still puts seven kilograms (gander) and six kilograms (goose) on the scale. Their carriage is slightly upright, giving a stately impression. The head does not have a dewlap and the belly has no stern. The head is of medium length with a flat forehead and sits on an upright-carried neck. They occur only in the white color variety.

The very attractive Sebastopol goose comes from southern Europe and apparently occurred as a mutation in ordinary indigenous geese. The German name "Lockengans" refers to the main characteristic of the breed, the formation of curls, which are elongated, spiral feathers. These often unusually long feathers have a stiff shaft that extends only two to three centimeters beyond the skin. From that point on, the feather is soft, flexible, and split into individual threads. The wide feather webs lose their cohesion and rotate slightly, causing the curl to develop. In some birds only the back feathers are curled; in others the entire plumage, with the exception of the breast and neck, is curled. The entire plumage is very soft, and the head is relatively small. The body produces a full and broad impression but is nevertheless short. The plumage

The Helmet Guineafowl is the source of domestic guinea fowl.

is pure white. With five to six and four-and-a-half to five kilograms respectively, gander and goose are quite light. The white eggs weigh about 120 grams. The laying capacity is about 25 to 40 eggs.

While the geese presented so far all go back to the Greylag Goose, in the Celler goose the Swan Goose as well as the Greylag Goose are ancestral forms. Their direct descent is based on brown and brown-pied indigenous geese. Their historical development occurred in the vicinity of Celle (Lower Saxony). With five-and-a-half to six-and-a-half kilograms in the gander and four to six kilograms in the goose, it is comparable to the Sebastopol with respect to weight. In type it is a medium size, mobile goose with a well-rounded breast and a single-lobed stern. The carriage is slightly upright, and the head is robust and resembles to some extent the long head of the Swan Goose. The plumage must be light leather-brown and merge into white at the belly. The leather-brown "mantle plumage" (body plumage) has a narrow light edging.

The Chinese goose, which was bred in China from the Swan Goose, came to Europe in the eighteenth century. At five or four kilograms — depending on the sex — and tight plumage, this goose gives the impression of being small. Its demands are relatively few; it does, however, show a special liking for greenfood. Owing to its vitality, the fertility of the eggs leaves nothing to be desired. In several clutches it produces an annual egg production of about 50 eggs. Typical of the Chinese goose is the trumpetlike voice, which is somewhat higher pitched in the gander than in the goose. An external characteristic of the breed is the elongated head with the medium-length bill, over the base of which there is a semicircular knob. With increasing age, the knob also increases in size. The neck is slender and curved like a swan's. It carries the elongated body with the broad shoulders upright; the hindparts are curved upward. It occurs in the colors gray and pure white. In the gray color variety, the bill and knob are black, being reddish-yellow in the white. The Chinese goose's gray plumage often has a brownish appearance, while the shoulders, wings, and thighs have a cream-white edging. The front of the neck and the upper breast are whitish-fawn. The breast exhibits a fawn-brown coloration. A dark-brown stripe ("eel stripe") extends from the occiput through the nape to the shoulders. Belly and hindparts are white; the gray tail feathers have white borders.

As in the Celler goose, the blood of Greylags and Swan Geese flows through the veins of the Steinbacher Game goose. In the last third of the nineteenth century it was bred in Thuringia as a fighting goose. In contests, the ganders went at each other with beating wings, biting until one of the rivals was no longer able to fight. Today the Steinbacher Game is kept only as a show bird, with a weight of six to seven kilograms in the gander and five to six in the goose. The egg production, at about 20 eggs, is quite meager. The long bill, which is slightly curved at the base and exhibits black "teeth" (horny, tooth-like protuberances on the ridge of the lower mandible), is characteristic. These geese move their robust, medium-size bodies agilely and proudly, very much in the manner of a gladiator. The body carriage is slightly sloping. A slight stern formation — in particular in adult females — is permitted. The plumage is tight, which gives the breed a very elegant appearance. The color varieties light-blue and gray are recognized. The head, neck, breast, back, and bill are always colored exactly like the gray tail feathers, while the shoulder, wing, and thigh feathers are edged with white. Belly and rump are white.

Ducks

Breeds Suited for Meat Production

The Aylesbury duck is a massive but not plump-looking bird. In England they were bred for meat production. Accordingly, they fatten easily and can be slaughtered early because they mature early. Additional positive traits are their robustness and vitality. The weight is about three and a half kilograms in the drake, while the duck is about half a kilogram lighter. Although the Aylesbury duck belongs to the meat ducks, the meat of which is very tasty, it still lays 60 to 100 eggs (with a weight of 80 grams) a year. The shell color varies between white and green shades. The plumage and skin color is pure white. The deep, broad, and long body is carried horizontally by the Aylesbury duck. The breast is full and well formed, as is the belly, which does not hang. The long, relatively narrow head has pronounced cheeks, which, however, do not cause the head to appear fat. The pale, pink-colored bill and the dark eyes contrast very well with the white plumage. The coarse-boned, dark-yellow legs become more and more orange with increasing age.

The Rouen duck exhibits a pleasant, quiet disposition and is very hardy and robust. Like the Aylesbury duck, this massive bird lays 70 to 90 eggs a year, but it does not begin laying until late. As a good meat duck, the drake produces three and a half and the duck three kilograms of live weight, which can be greatly increased if given a pure fattening diet. Their appearance is similar to that of the wild Mallard, their original ancestors. Drake and duck exhibit different plumage colorations, exactly in the manner of the wild ducks. The trunk is long, broad, and deep. It embodies a so-called square style. The keel forms a fold of skin running from the protruding breast to the hindparts. The back exhibits a slight arching. The slim head has a small forehead and does not exhibit "cheeks" (slight fullness of feathers on the cheeks). The color of the bill is different in the two sexes. In the drake it is olive-green and brownish-yellow, while the duck's bill has deposits of black pigment. The orange-red legs and toes have black claws. The 80 gram egg of this breed from northwestern France varies in coloration between white, greenish, and bluish. The Rouen duck reached England from France, and was bred there for color and size before it came to Germany in the second half of the nineteenth century. The Rouen duck is highly recommended to breeders who cannot offer their ducks an opportunity to swim.

A widely distributed meat duck in Germany is the Pekin duck. This breed exists in two breeding strains. The American Pekin duck only lifts its body slightly above the ground, while the German Pekin duck has a very upright carriage. Because it lays 100 to 130 eggs weighing between 70 and 80 grams, it became a first-class production bird. Even the ducklings are resistant to harsh weather conditions. In temperament it is somewhat timid and often quite noisy. The Pekin duck comes from northern China and reached England and the United States first.

In the United States, the American Pekin duck was developed with the aid of the Aylesbury duck from the Chinese ancestral form. It arrived in Germany at the beginning of the twentieth century. It carries its robust body slightly raised. The pure-white plumage lies close to the body. The fairly long skull shows a slight forehead. The long, light-yellow to orange-colored bill and the dark eye stand out from the white plumage. The drake weighs three kilograms, the duck two and a half.

The German Pekin duck was bred in Germany in the 1870s. Drake and duck are always around a half kilogram heavier than their American relatives. The uprightly carried, massive body resembles a square and is almost twice as long as it is deep and wide. Bill and head exhibit a short, broad and high form, respectively. A full and heavy breast without a "keel" (breast-bone process) and the full, broad belly, as well as the similar hindparts give the German Pekin duck the complete attributes of the meat duck. The white plumage is suffused with yellow.

The Cayuga duck is striking because of its black, metallic-green, gleaming plumage. Unfattened it reaches, depending on sex, two-and-a-half to three kilograms, yet the flesh tastes gamy. The skin is, so to say, consumer-friendly white. With 70 to 100 white to greenish eggs it also exhibits good egg production. It was bred near Lake Cayuga in New York State. It reached Germany in the 1870s. This medium-size duck carries its rounded body almost horizontally. On the medium-size, nicely curved neck sits a small, oval head with a flat forehead. The bill has an olive-green pigmentation.

The Pomeranian duck is known in various European regions. It often carries as a synonym the name Swedish duck. In Germany, it was named for the region of origin. With 90 to 120 round, 70 gram, usually colored eggs, this meat duck also exhibits good laying production. The drake weighs three kilograms, the duck weighs about a half a kilogram less. The youngsters reach slaughtering age at ten weeks. The Pomeranian duck is very hardy; it much prefers to have the opportunity to swim. The Pekin duck has displaced it as a commercial breed in many areas. The total impression is that of a heavy domestic duck shape with a long, broad, and deep body, which despite its size does not appear plump and is carried horizontally. The full and protruding breast, which merges into the broad, deep belly, does not exhibit a keel. The head has a flat forehead and a long, broad bill. The Pomeranian duck occurs in blue and black color varieties. A white patch, which is also called the "bib," stands out from the front of the neck and the crop.

The Saxon Duck was bred in Saxony during the 1950s from crosses of Pekin, Rouen, and Pomeranian ducks, and was recognized in 1958. This robust indigenous duck with a long, broad body and an almost horizontal carriage is marked by pronounced efficiency. At ten weeks, this fast-growing and full-fleshed duck is ready for slaughter. In addition, it surprises with a relatively good laying production. The 80 gram eggs are white. This breed only occurs in the blue-yellow color variety. In the drake the neck and head are pigeon blue up to the closed, white neck ring. The lower neck, breast, bend of the wing, and shoulders exhibit a rust-red coloration, while the breast is lightly edged in silver. The lower back and rump are pigeon blue. The tail and the wing feathers exhibit a mealy color component. The belly and wings show a blue-gray color. In the duck the head, neck, and breast are a deep pea-yellow, whereby the head has a white eye-stripe. The open neck ring is required. The back is a light pea-yellow. Rump and tail are light blue; the wing coverts are cream-colored with a blue tinge.

The Gibsheimer duck represents a recent duck breed, which was not recognized until 1963. The Pekin, Orpington, and Saxony ducks contributed to it. This broad and long, more than medium-size indigenous duck with slightly upright carriage has good fattening ability and an abundant amount of meat. The drake weighs three kilograms, the duck two and one half kilograms. The yellowish to green egg has a weight of 70 grams. So far, this duck has, however, not yet become established as a popular breed, although the blue-gray coloration of the plumage is something out of the ordinary.

Another recent breed, which was developed in the 1970s, is the Altrheiner Magpie duck. In form it resembles the Gimbsheimer duck and is also similar to that breed in weight. The pure-white ground color is interspersed with black pied patches. The pied markings extend over the crown of the head, the shoulder plumage — whereby the heart-shaped markings overlap onto the wings — and on the upper part of the tail.

The Muscovy is the only domestic duck that did not stem from the Mallard, but instead from the Muscovy Duck of South America, which was already domesticated there in the sixteenth century. Typical of the breed is the bare, warty face, the bill knob, the long neck, and the lack of "curls" in the drake (seasonal secondary sex characteristic on the drake's tail). The four kilogram male exceeds the duck's weight by about one kilogram, but can, however, also easily reach the five-kilogram mark. The flesh is, to be sure, darker than in other ducks, but, on the other hand, is very juicy. The laying production is a maximum of 100 eggs; as a rule, however, the number is significantly smaller, since the ducks incubate dependably. The Muscovy does not absolutely require a water basin for swimming, and, because of its virtually silent voice, does not make itself disagreeably conspicuous. Some birds show a tendency to fly, for which reason clipping the wings of the occasional one is appropriate. The trunk exhibits an elongated and very broad shape with a horizontal carriage. The crown feathers are long and slightly stiff. The bill has a tooth on the tip. The eye region is free of feathers and warty. The head has an S-shaped bend. The color varieties wild-colored, blue wild-colored, blue, black, white, pied, and pearl-gray are recognized.

Laying Breeds

The Runner duck represents a laying breed of first-class quality. Its laying production reaches 200 eggs. The 65 gram egg is usually white, but is greenish in the darker color varieties. Laying begins in the fall and continues over the winter into spring. Egg production is promoted by a large run. With a weight of two kilograms for the drake and one and three quarters for the duck, this breed is relatively light. Since Runners are very hardy, they are also excellently adapted for our latitudes. In addition, they can be kept as a substitute for chickens in those places where the morning cock call is considered a disturbance. Runners, because of their fine-boned legs, do, however, show a tendency toward leg ailments, which has a negative effect on vitality and production. The Runner comes from Southeast Asia and was imported into Germany about 100 years ago. The overall impression is that of an upright running, streamlined, and symmetrically formed duck. Unfortunately, it is very lively and therefore nervous. The narrow, long-faced, and angled head with a flat forehead merges with a sharp bend into the wine-bottle-shaped neck. This again fits in harmoniously with the cylindrical body. The back is well rounded. The tail continues the line of the back. The wings fit into the well-rounded torso. Many color varieties occur: there are wild-colored, trout-colored, white, black, brown, blue, pea-yellow, and fawn white-pied Runner ducks.

The Campbell duck was bred in England. Because of the strong laying production of 180 to 200 white to greenish eggs, this breed soon became very popular. From time to time some individuals even exceed the 200-egg threshold. The minimum egg weight of 65 grams is often exceeded. In addition, the 2 to 2½ kilogram Campbell duck is fast growing and easy to raise. This is a lightly built, slender duck with a somewhat upright carriage. The trunk is fairly long and well rounded, whereby the upper and lower lines of the body are parallel. These ducks are found in khaki and white. In the first color it is a matter of a brown color combined with a tinge of red. In this duck, the wild markings can be seen dully through the ground color. In the white color variety, a pure white and white skin are demanded.

The Streicher duck comes from crossing Campbell and Runner ducks. In its English homeland it is called the "Welsh Harlequin." In the 1920s it entered Germany through Denmark. On account of its ancestors, the laying production is very high; the 65 gram egg has a white coloration. In form, the lightly built Streicher duck with its weight of two to two and-one-half kilograms inclines toward the Campbell duck. It is bred in the lightest wild color, the silver wild-colored shade. In the drake the ground color is silvery cream-white. The breast, base of the neck, nape, and shoulders are red-brown with silver-white edging. The lower back exhibits a silver-gray color and dark spots, whereby each feather is edged. The brownish-black head is an iridescent green and carries an unclosed neck ring on the rear. In the duck the ground color is yellowish-white, while the breast, base of the neck, nape, and back are lightly striped in brown. The lower back exhibits dark spots with white edging. The head shows a brownish-yellow color component.

In England, Sir William Cook bred the Orpington duck from the Aylesbury and Pomeranian ducks toward the end of the nineteenth century. With 150 to 180 approximately 65 gram eggs, production is very high; at the same time they are recommended because of their juicy, tender meat. Thus, in a real sense they are a dual-purpose breed. This duck exhibits a robust constitution and rapid growth. It gets by with relatively little feed and a small enclosure. On the other hand, it bathes very readily, for which reason a bathing area is indispensable. It carries the cylindrical body slightly elevated. The belly and hindparts are fully formed but do not touch the ground. The long and narrow head has a flat forehead. The plumage of this duck, which weighs two-and-one-half to three kilograms, is a uniform leather-yellow. Occasionally, the head and neck of the drake is chocolate colored.

Ornamental Breeds

The domesticated Mallard resembles the wild Mallard very closely. It originated at the beginning of the twentieth century from crosses between Mallards and domestic ducks, in the course of which it retained the ability to fly. This breed, developed in Germany, favors nesting sites placed high up. Escaped domesticated Mallards transmit their genes to wild Mallards, which is why this species is no longer genetically pure, as has been documented externally through varying plumage coloration. At one and one half kilograms (drake) and one and one quarter kilograms (duck), domesti-

The upright stance marks the Runner duck—in this instance, the penciled variety.

cated Mallards are relatively light, but somewhat larger than their wild ancestors. The boat-shaped trunk exhibits a flat bottom line and an only slightly curved top line. In contrast to the Mallard, the fairly long head only shows a small forehead. From time to time a spherical, closed crest adorns the head. The neck is slightly curved. Domesticated Mallards are ready to slaughter within the space of eight to ten weeks, and the tender flesh has a delicious flavor. Their feathers are well suited to filling pillows. As furnishings for the enclosure, they prefer to have a place to hide and to swim. Under all circumstances the enclosure should be covered above, so that flying away is impossible. Only in this way can a loss of birds and a further adulteration of the wild Mallard population be prevented.

The hardy Crested duck represents a hundred-year-old mutated form of an indigenous duck that used to occur in Germany and the Netherlands. With a weight of 2 to 2½ kilograms, depending on sex, and 120 eggs a year, this breed, apart from its unqualified beauty, also has commercial value. As its name indicates, it carries a spherical, standing, closed crest on its occiput. The compressed, meaty body exhibits a horizontal carriage. The curved neck merges into a longish, round head with protruding cheeks. The breast and belly are fully fleshed. The Crested duck occurs in all indigenous duck color varieties, and in the colored varieties the crest is usually a lighter color than the remaining plumage.

The home of the black, lustrous, emerald-green East India duck is Brazil; it is, however, descended from — as is true of almost all domestic ducks — the Mallard. Since the Mallard is not found in South America, however, its true origin remains lost in obscurity. The drake weighs one kilogram, the duck three-quarters of a kilogram. Along with the Call duck discussed below, the East India duck represents the lightest breeds. They carry the fairly long, rounded trunk almost horizontally. The lightly elevated breast is well rounded. The legs and bill are suited to the dark plumage coloration. At times their laying production rises to 80 eggs; for the most part, however, it remains below this figure.

The Call duck may not weigh much more than one kilogram. Its beginnings are unknown. The reason for its name is that it was widely used to lure wild Mallards which the hunter could then easily shoot. The breed was recognized in Germany in 1943. In egg production it is comparable to the East India duck. The deeply carried, well-rounded body gives the impression of being very short and small. The fully developed cheeks of the rounded head with its pronounced forehead is a characteristic of the breed. As with the domesticated Mallard, a crest can sit on the occiput. The neck exhibits a slight bend. Like the domesticated Mallard, it occurs in numerous color varieties. Mainly because of the multitude of colors and the ease of care, the Call Duck is very popular.

Turkeys

The common turkey only occurs in a single breed, but, on the other hand, there are numerous color varieties. A size designation is coupled with each color variety. Accordingly, turkeys are separated into heavy, medium-weight, and light varieties. Correspondingly, heavy turkeys weigh 9 to 12 kilograms as young toms and 12 to 15 kilograms as old toms, 6 to 7 kilograms as young hens and 6 to 8 kilograms as old hens. Medium-weight turkeys are 8 to 10 kilograms as young toms, 10 to 12 kilograms as old toms, 5 to 6 kilograms as young hens, and 6 to 7 kilograms as old hens. In the light color varieties, the young tom weighs 6 to 7 kilograms, the old tom 7 to 8 kilograms, and the young and old hens 4 to 5 kilograms. The minimum weight of the egg is 70 grams in all color varieties. The colors bronze, black-winged, and white are represented among the heavy turkeys; black, red-winged, and bourbon among the medium-weight turkeys; and blue, red, buff, copper, and royal palm among the light turkeys.

The common turkey exhibits a large and powerful appearance, in which the elongated trunk is especially broad over the shoulders and tapers in a sloping line toward the tail. The bare head, which is covered with red caruncles, with its blue to lively sky-blue coloration, is particularly striking. From the base of the forehead over the beak hangs a dewlap, especially pronounced in the tom, which lengthens when the turkey is excited. The tom's head has no feathers at all; in the hen, the crown exhibits a sparse growth of feathers. By this means it is possible to distinguish between young hens and toms early on. The neck is slightly curved, and in its upper region it has a conspicuous, bare area of skin of a reddish-blue color. A beard of varying prominence grows on the breast area of toms as well as on some old hens. The turkey carries its long, broad wings high and close to the body. The long tail, which is carried low, is closed and is raised like a fan in the male. The turkey's legs should be very long, which imparts a certain amount of elegance.

In the bronze color the breast, neck, shoulders, and bend of wing have a black ground color with an intense bronze sheen, which shimmers in the colors of the rainbow. The back feathers exhibit a 1–2-centimeter-wide, gold to violet-red, lustrous bronze band, which has a narrow black stripe and a broad chestnut-brown stripe on the tip of the feather. The grayish-white wing feathers show uniform, distinct black diagonal bands. The tail feathers have a 1–2-centimeter-wide bronze band on the tips of the feathers, which is set off by a sand-colored to golden brown stripe.

White turkeys are pure white; only the beard is black. A velvety, lustrous black marks the black color variety. A uniform dark or light blue is required of the blue color variety, in which the plumage may be dotted with black. In the uniformly colored red color variety, white wing tips are allowed. A rich ocher yellow with a deep yellow underflue is required in the yellow color variety. The bright, rich copper color should have a lively sheen in the copper color variety. Frequently, a blackish-blue feather edging occurs in this variety.

In the royal palm variety the primary color is white, whereby each feather has a black edging, which is again enclosed by an edging of white. Depending on sex and area of plumage, the edging is more or less well defined.

In the red-winged color variety, the ground color of the plumage is an orange-red to olive-green, lustrous dark leather-brown color, in which each feather is edged in black in the tom and in a sand-brown to reddish-brown color in the hen. The primaries are grayish-white with rich black dots. The secondaries exhibit a lively pinkish-red color with fine black dots. Because of the iridescence of the deposited gold pigment, the red-winged turkey is very attractive.

In the bourbon tom all feathers are dark brownish-red with the exception of the black-edged neck feathers; this edging is lacking in the hen. The wing feathers are white as is the tail, and the tail exhibits a red, white-edged diagonal stripe on the feather tips. In the hen, the breast plumage has small, white edging.

The black-winged turkey has a deep black ground color, which, depending on the way the light falls, is an iridescent dark bronze to bright green. From the shoulder to the tip of the tail, each feather carries a so-called gold plating. The edging of the feather tips on the tom's back are black, lustrous green in the hen, on the tail covert feathers brown, and on the tail feathers light brown. The wing feathers have a black coloration, of which the secondaries show a narrow white edging.

Guinea Fowl

Guinea fowl are principally kept for their tasty flesh and delicious eggs. In addition, they are a popular breed of poultry because of their unusual appearance, although their screeching usually has a disturbing effect.

The domestic form of the guinea fowl is heavier than the wild form; surprisingly, the cock, at 1.6 to 2 kilograms, weighs less than the hen at 1.8 to 2.5 kilograms. The yellowish to brown eggs weigh from 40 to 45 grams.

The hen has a fleshy, elevated breast, while it is more pointed in the cock. The back shows an arching and slopes to the rear. The tail, consisting of 16 feathers, is carried low by the guinea fowl. The most interesting feature is the head. It is short and broad, and a horny, helmetlike, triangular brown structure ("helmet") with the tip pointing to the rear is placed on the crown. The dark-brown eye and the sharply curved, orange-red beak increase the attractiveness of the bare, bluish-white face. The wattles are red with a white patch.

A characteristic for distinguishing the sexes is the helmet, which is pointed toward the rear in the male and is blunt in the female. The cock's wattles are more prominent, and, in contrast to the hen, they are not smooth. The color varieties blue, pearl-gray, lavender blue, dundotte, white, violet, azure blue, lavender blue with reduced pearling, and dundotte with reduced pearling are known. In the blues, the entire ground color is gray-black; each feather is indigo blue, however, when examined individually, and many white dots ("pearls") are present on each feather. On the back, rump, and tail, the pearls are ringed in black. In the pearl-grays, the feather has a light gray-blue ground color, and the pearling is surrounded by blue instead of black. The lavender blues show a subdued light-blue ground color. The dundotte-colored birds have a fawn-yellow coloration with rich buff, enclosed pearling. Sexually determined, the cock's ground color is paler. The whites occur in a cream-white ground color with lustrous, silver-white pearling. In the violets, the plumage color is black-violet, whereby each feather is edged in deep black. The plumage on the flanks and the wing feathers exhibit a fine white pearling or striping. The azure blues are gray-blue with violet overtones; the cock has a sex-linked darker ground color. The pearling corresponds to that of the violet guinea fowl. In the lavender blues with reduced pearling, the ground color is a subdued light blue with a bluish film. The color of the markings corresponds to that of the violets. The dundotte variety with reduced pearling have the same ground color as the dundotte variety but have the pearling of the violet guinea fowl.

FEEDING

Body Composition and Energy Provision

The bodies of our poultry are, like all other organisms, made up of individual cells, to which all life and metabolic processes are bound. If one analyzes the animal organism with regard to its basic chemical elements, then it can be seen that it is composed of carbon, hydrogen, oxygen, phosphorous, calcium, magnesium, iron, sodium, sulfur, iodine and many other elements.

So that the organism can grow and develop, it must constantly take in these raw materials. By means of numerous chemical steps the body incorporates the substances it has taken in. In so doing, innumerable synthesis, conversion, and degradation reactions take place. So that the chick develops optimally into a hen, that the adult bird molts satisfactorily, or that a high laying production is achieved, a diet with high-quality foods that provides the body with the right nutriments at exactly the right time is necessary.

For so-called synthetic metabolism — that is, growth — principally proteins (nitrogenous foodstuffs) are needed. Carbohydrates and fats, on the other hand, are more important in supporting energy metabolism; that is, providing energy for managing everyday demands. In their totality and in their action these nutriments strongly support and affect one another, so that one can speak of an exchange system on the one hand between the nutriments and, on the other hand, between the body and the nutriments.

Basic Components of Foods

One recognizes various basic components from which a food is made. The diagram on the following page should make this clear.

If one applies the general model to a specific food, then this means that the components of a barleycorn, for example, can be divided into the dry mass and the water fraction. In barleycorn, the water fraction and the dry mass are found in the proportion 14:86. The proportion of water is relatively small in barley, since a green plant, for example, has a water content of up to 98%. Because the barleycorn — like all other seeds — serves for reproduction, however, its water fraction is greatly reduced, because in this way the life processes are slowed down and thus a longer dormant period (storage) is possible. The carrot, a metabolically active plant, has exactly the reverse water–dry-mass proportion as the barleycorn, namely 86:14.

If one heats the particular foodstuff, all water evaporates and the dry substance remains. This consists of organic, combustible constituents and inorganic, incombustible components. Of the barleycorn's 86% dry mass, 82.7% is organic and 3.3% is inorganic substances. The latter is also called "raw ash." Raw ash can be divided into primary elements such as calcium, phosphate, potassium, magnesium, sodium, sulfur, chlorine, as well as into trace elements such as iron, copper, manganese, iodine, fluorine, and many others. The organic mass is divided into the complex of organic building blocks and organically active agents. Among the latter, we recognize vitamins, and also enzymes, hormones, antibiotics, and the like. They play a very active role — even if in only very small concentrations — in the body's metabolic processes. The organic building blocks can be divided further into the nitrogenous and the nonnitrogenous substances. The nitrogenous substances break down into the proteins and the nonproteinlike components of amides. Together they provide the previously mentioned basic substance for building

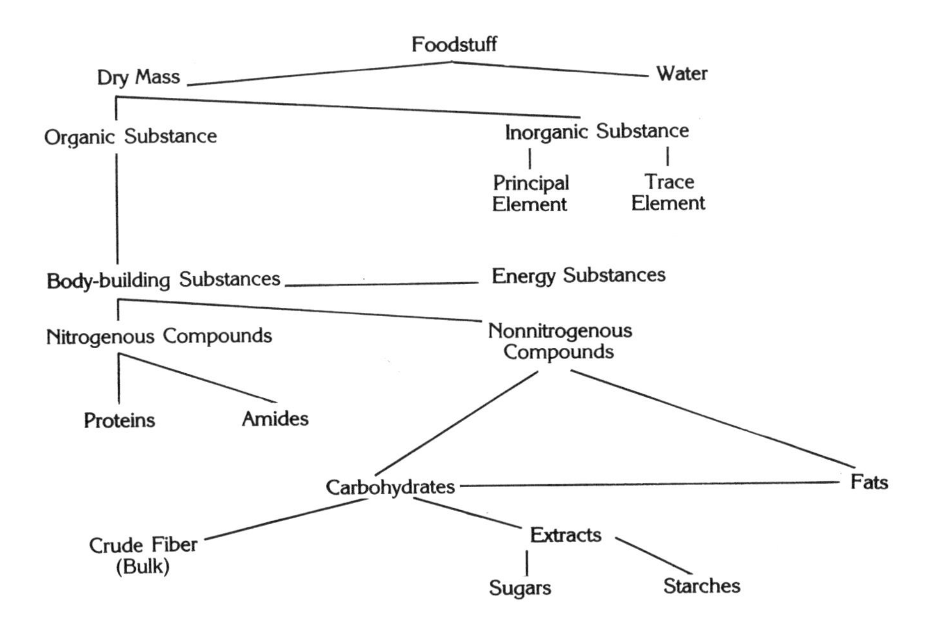

the body. So that they can be correctly incorporated into the existing body mass, a complex interplay with numerous substances is necessary, above all with the organically active agents. The complex of nonnitrogenous compounds, to which fats and carbohydrates belong, represents the source of energy. Carbohydrates can be further divided into a raw-material component, which has no actual nutritional value, but only fulfills a ballast function, and into the extract component. The extract component consists of easily usable sugars and starch, which — like the fats — represents a stored energy source, to be drawn on during times of need.

Proteins

Protein is the primary constituent of every cell, and is the medium for numerous life processes. It is made up of smaller building blocks, the so-called amino acids. Of the 23 amino acids commonly found in animals, about 12, depending on the species of animal, are essential, which means as much as "necessary for life."

In the case of poultry, some of these essential amino acids are methionine, lysine, threonine, and glycine. In contrast to the semi-essential (vital) and nonessential amino acids, poultry cannot manufacture the essential amino acids in their own bodies, so they must obtain them from the outside.

Amides

Amides are principally active in the degradation, conversion, and synthesis of protein. Amides and proteins are often combined under the often-used concept of "raw protein." A high amide content is found chiefly in young plant components and in seeds.

Carbohydrates

The carbohydrate category includes many kinds of sugars, which, depending on their chemical composition, are classified as simple, complex, and compound sugars. Glucose is a familiar simple sugar, sucrose is an example of a compound sugar, and starch, insulin, and glycogen are complex sugars. So that the sugar can be usable as a source of energy for muscles or to produce heat, it must be present in the form of a simple sugar. Complex sugars serve as storage substances; for example, glycogen is stored in the liver.

Cellulose — also a complex sugar — is classified as crude fiber. Cellulose is characterized by being very difficult to digest, and is primarily decomposed in small amounts in the caecum, where it remains for a short time. Although it does not have great nutritional value, it is of significance for optimal digestive processes.

Fats

Fats produce heat in the body, which is why they are also called "heat components." As deposited fat, they are stored in, for example, the subcutaneous tissue and on the viscera. Fats provide the greatest amount of energy per weight. Experiments have shown that fats produce 2.3 times more heat than the same amount of carbohydrate. Besides the deposited fat, poultry also use the so-called organic fat, which is constantly synthesized, converted, and degraded. Unsaturated fatty acids, in particular, take part in these metabolic processes, which are vital to the organism and required in a constant supply. For this reason, every feed mixture must contain a certain amount of fat or a certain mixture of fat. In addition, body fat is the vehicle for the fat-soluble vitamins A, D, E, F, and K.

Fatlike compounds, the so-called lipids, effect the transportation of fat in the body, promote the digestion and resorption of fat, and, in the form of carotene, serve as provitamin A. They also make up the wax of the uropygial gland, which serves to oil the plumage.

Minerals

Mineral elements, which can be divided into primary and trace elements, are very important for the birds' health. Since the primary as well as the trace elements are essential for the function of animal metabolic processes, a mineral deficiency leads to growth and developmental defects in the youngster and to a decrease in production in the adult. Because minerals, just as vitamins, are of great significance for nutrition in poultry, they will now be described at length.

Calcium is of great importance for the formation of the skeleton. The formation already takes place in the chicken egg during embryonic development. Over an embryonic cell layer, the growing embryo decomposes part of the calcium of the eggshell and integrates it into its skeletal system. By means of this mechanism, sufficient calcium for bone formation is available to the embryo, and the calcium shell simultaneously becomes easier to crack during hatching. In addition, calcium ensures the functional efficiency of the body's cells. At the same time, the mineral brings about adequate cell permeability for fats and proteins. To a great extent, the blood is dependent on calcium, since it supports blood clotting. Above all because of the formation of eggshell, calcium stands at the forefront of the essential mineral elements. Normally, the eggshell is made up of 93% calcium carbonate. With every egg laid, the bird's body secretes five to six grams of calcium from the blood. Since the chicken is incapable of storing large amounts of calcium, the deficit must be balanced through the diet.

Because of this fact, the breeder allows his chickens to take in as much crushed oyster shells and edible calcium as they want, over and above the eggshells from his own chicken flock. In small doses, one can also feed calcium chloride. For this purpose, the breeder dissolves 0.2 grams of calcium chloride per bird per day in the drinking water. Chlorine poisoning can occur in higher concentrations. Meat, fish, and bone meals also have a high calcium content. In addition, some veterinary pharmaceutical companies also offer calcium preparations. Apart from ensuring the mineral supply as such, the proportion of calcium to protein must also be in balance. If growing youngsters receive a feed that has a high percentage of protein, then their muscle mass grows significantly faster than the skeletal system can keep up with. As a result, the too high body weight presses on the too weak skeletal system, and skeletal deformities — especially crooked toes — occur.

If the supply of calcium remains far below the minimum requirement, the result can be brittleness or weakening of the bone, above all when D-avitaminosis (vitamin-D deficiency disease) is added to the calcium deficiency. But a superabundant supply of vitamin D also leads to damage. In this case, too high a concentration of calcium occurs. If youngsters grow too fast, then the pressure of the perch on the sternum is so great that it leaves an imprint (depression). In this case, the unsatisfactory formation of the sternum is of course not hereditary and is not given minus points at poultry exhibitions.

Besides this more anatomical point of view, calcium also plays a very large role in nerve and muscle metabolism. Along with potassium and sodium, calcium plays a significant role in the transmission of nerve impulses. It also produces muscular contraction and has a regulatory effect on the heart action. Chronic calcium deficiencies result in retardation of growth, rachitic symptoms, lameness of the extremities, loss of appetite, and enervation.

Another essential primary element is phosphorous. Together with other minerals, it regulates metabolism — above all, that of carbohydrates and fats — and represents a critical component for providing energy in the body. In addition, phosphorous is found in numerous substances in the body, such as the cell nucleus, enzymes, and hormones. In the formation of the continually renewed body cells and bone, phosphorous plays a role that should not be underestimated. In the nucleic acids — genetic chemical compounds — it is significant in the transmission of genetic information. Phosphoproteins provide nourishment for the developing embryo in the egg.

Besides the importance of phosphorous in itself, its relation to calcium is also relevant. If there is a surplus of calcium in the blood, phosphate is taken from the bones, so that softening of the bones can occur. If too little calcium is present in the blood, a softening of the bones also takes place, which, however, proceeds distinctly more slowly than with a calcium surplus. In a phosphate deficiency, retardation of growth and reduction in production, as well as anemia and general weakness occur.

The proportion of phosphorous to calcium should be 1:1.5 in youngsters. In laying hens, the ratio of 1:3 is favorable, and brooding hens should have a ratio of 1:2. The same ratio is also important for optimal sperm production in the breeding cock. If an incorrect ratio of these substance is present, in addition to osteomalacia (softening of the bone), bone deformities and swelling of the joints can appear. Additionally, vitamin D no longer functions optimally, through which the osteomalacia advances particularly rapidly. Vitamin A also plays a role in this regulatory mechanism. In light of this mineral-vitamin metabolic complex, the interconnection of the metabolico-physiological relationships become apparent. Rye and wheat bran, cereal grains, and fish, meat, and bone meals are rich in phosphorous. The phosphorous component of cereal grains is, however, in part indigestible.

Sodium and chlorine occur predominantly in the body fluids of the organism, and regulate the transportation of the fluids through the body's cells. Chlorine also plays an important role in digestion in the gut. Chlorine and sodium are usually provided in the form of table salt. Yet the chicken should receive at most one gram a day. Two to three grams a day cause irritation of the internal organs, and 4½ grams is enough to poison the chicken. If the food has the proper concentration of sodium and chlorine, it will increase the appetite since it makes the food appetizing. Sodium fosters the proper functioning of the nervous system and regulates the acid-base balance. Abundant amounts of sodium chloride are found in shrimp.

To some extent, potassium is the counterpart of sodium in the body fluids, influencing the function of the heart muscle. In addition, it has an effect on the skeletal musculature. It also fulfills an important function in the nervous system. Moreover, it is found in virtually all bodily substances. Since potassium is always found in greenfood, a deficiency seldom occurs. Typical deficiency conditions would cause a decrease in production in adult birds and poor growth in chicks. Extreme cases can even lead to death.

Magnesium is an essential factor for tissue growth. It is also active in protein conversion and bone synthesis. At the same time, it is a component of the enzyme that activates sugar molecules. In addition, it binds surplus acids. A deficiency leads to blood-vessel dilation, overexcitability, kidney damage, convulsions, stoppage of growth, and slow feathering. Finally, death can also result. Magnesium also plays a critical role in the so-called seminal fluidity of the ejaculate, where, together with calcium, sodium, potassium, and chlorine, it promotes fertility in the cock. Greenfood, especially turnip greens, is rich in magnesium, but an abundance of it is also found in bran.

Sulfur is a building block of protein and several amino acids. These amino acids are essential for energy-related metabolic processes. Sulfur is of particular significance in the molt. An increased sulfur supply is necessary for the formation of feathers. Feeding elementary sulfur is unsuitable, since the bird cannot assimilate it. It is, however, effective for a light worm infestation. Under certain circumstances giving it can cause poisoning. It can be used by the body when given in the amino acids methionine and cysteine, and the vitamins B_1 and biotin. High doses of sulfur should be given in conjunction with vitamin D, so that softening of the bones does not take place.

One of the most important trace elements is iron. It is a significant component of the red blood pigment hemoglobin. In addition, it is also found in ferritin, an albuminous substance which is of significance in iron resorption. It is also active in energy production, as well as being a constituent of some vitamins. With iron deficiency, an iron-deficiency anemia occurs, and the birds also suffer from the symptoms of general weakness. Iron deficiency can often be traced to one-sided feeding of food rich in carbohydrates or excessive feeding of milk. Iron deficiency also develops quickly with parasite infestations or other illnesses. Iron can be provided through greenfood and meat and fish meal. The often-used rusty nail in the drinking water is not useful for preventing iron deficiency. The providing of green (iron) vitriol in the drinking water is said to have a positive effect on blood-building and metabolic processes in the animal organism. The opposite opinion, however, also exists: that green vitriol, besides having a slight constipating effect for diarrhea, has no particular significance in physiology or metabolism.

Cobalt is a component of vitamin B_{12}, and plays a role in blood formation. In metabolism, it regulates enzyme action in part. Cobalt can be provided through cobalt sulfate or vitamin B_{12}.

Zinc is a component of insulin which, together with glucagon of the pancreas, regulates the blood-sugar balance. In addition, it decisively affects the activity of various enzymes. It is also important in the formation of horny structures like the toenails. In the organism, zinc is found principally in the brain, glands, and reproductive organs. It plays an important role in growth and in the function of the liver and kidneys. A zinc deficiency primarily affects reproduction. It is easily provided in the form of zinc sulfate.

Copper is present in virtually all organs. It guarantees healthy development and normal growth. It is primarily found in compounds containing the elements iron and cobalt. For example, with a copper deficiency the iron in the organism does not take full effect. Copper apparently also plays a role in the formation of blood. It can be provided in greenfood, meat and fish meal, as well as copper sulfate.

Manganese is a component of several enzymes and possibly takes part in blood and fat synthesis. Additionally, it plays a substantial role in the endocrine system. In hens it improves laying production, and in chicks it improves growth. Manganese also supports the hatching process in the egg. In a deficiency, many embryos die since they are unfit for life because of deformities. In hatched chicks, a deficiency becomes apparent in leg-joint disease. Wings and legs are often shortened. Eggs exhibit poor shell development. In general, heavy breeds show more susceptibility to manganese deficiency than do light breeds. The manganese requirement can be supplied through manganese sulfate.

Although iodine is only found in small quantities in the body fluids, it plays an important role in regard to physiology. For example, it is essential for the synthesis of the thyroid hormone. Iodine deficiency causes abnormal growth in embryos and chicks. In earlier times, the poultry breeder gave iodine, which was supposed to stimulate the ovaries, to improve laying production. This has since proved to be incorrect; instead, too much iodine causes a decrease in production.

Fluorine is a disputed trace element. In too high a dose it has a toxic (poisonous) effect. Silicon, which is principally found in grass, has a favorable effect on the development of connective tissue and on the elasticity of the skin.

Though mineral elements are present in prepared feeds like laying mash and laying-hen complete mash, from time to time the chicken's requirement exceeds this amount. Particularly with the mineral elements calcium, phosphorous, sodium, manganese, and zinc one often finds an insufficiency. For this reason, the breeder should offer his chickens a mineral mixture as a dietary supplement. The feed industry offers numerous preparations for this purpose. An increased and haphazard supplement of minerals, however, can also lead to a decrease in production, for which reason the mineral mixture should be fed in moderation. From plant sources, greenfoods of all kinds are relatively rich in minerals, particularly in calcium, phosphorous, magnesium, iron, manganese, and copper. From animal sources, bone, meat, and fish meals especially contain an abundance of minerals. Even the blood of slaughtered animals has a high mineral content.

Providing a so-called mineral block, which is enriched with vitamins, has also proved its worth. These are readily taken by chickens, turkeys, and guinea fowl, and cover at least a part of their mineral requirement.

Besides mineral elements, the birds also need small stones for optimal digestion. When kept in free runs, the birds usually find enough themselves; when kept in coops, they must be offered to the birds. Since birds do not have teeth, as is commonly known, the stones in the gizzard serve as substitutes for teeth, and help to grind up the food. After a time, the stones become rounded off because of abrasion and are expelled, which is why they must be taken in regularly. A chicken will have from 4 to 18 grams of gravel in its gizzard, a duck will have 10 grams, and a goose will have 30 grams. The stones should be given in the form of grit; but offering quartz sand fulfills the same purpose. Providing grit is essential when feeding seeds, whereas a grit supplement is not as necessary when using a complete feed.

Vitamins

The category of "vitamins" comprises specific organic substances that regulate and direct metabolic processes in the organism along orderly pathways. As a rule, the body cannot synthesize its own vitamins; therefore, they must be provided in the diet. The required amounts are very small but have a great effect on the organism. A few vitamins are transformed in the body into an active form through the conversion from an inactive precursor. Vitamins are principally found in greenfood but can also be found in meats because of previous storage. A lack of vitamins leads to symptoms of deficiency, so-called avitaminosis.

In chickens, ducks, geese, turkeys, and guinea fowl, under certain conditions (for example, after an illness; with early production of eggs for hatching, during the laying period) there can be a need for increased vitamin intake.

Based on where they occur, vitamins have hitherto been classified as either fat-soluble or water-soluble. They are still primarily designated with capital letters, although recently the American trend of listing vitamins under their chemical names has become more widespread. For this reason, in the interest of clarity, both designations are given.

Fat-soluble Vitamins

Vitamin A (retinol) promotes the development and growth of young poultry and the laying production of adult birds. It strengthens resistance to infectious illnesses and intensifies the metabolism of the skin and mucous membranes. It also works to prevent winter colds in poultry.

A retinol deficiency causes loss of appetite and retardation of growth. The egg yolk is often pale. The chicks' hatching ability diminishes, and the hatching date is delayed. Hatched chicks appear limp, feeble, and without vigor. Besides a pale beak and head as well as legs deficient in pigment, their entire constitution leaves something to be desired. Whitish-yellow deposits, which are easily removed, often form in the beak. The breeder often discovers eye inflammations or running eyes. Such susceptible and weak birds are often subject to infectious illnesses, particularly colds and intestinal illnesses, in the course of which symptoms of diarrhea appear.

Since vitamin A is stored in fairly large quantities in the body, with a deficiency the symptoms of illness do not appear until late. The vitamin is found in the form of its precursor (provitamin A) in carrots, greenfood, corn, milk, and animal fats in sufficient amounts. Actual retinol can only be obtained from the animal kingdom. It is found in the fat of most fishes, especially in cod-liver oil. The vitamin A content is given in International Units (I.U.) as well as in milligrams, whereby 0.3 milligrams is equivalent 1000 I.U.

The breeder can supply retinol particularly easily and in concentrated form as cod-liver oil; it is best given by way of pressed meal or seeds. For this purpose, the breeder fills a bucket with the appropriate feed and mixes it with cod-liver oil. After the bucket is covered with a lid, one shakes it so that the cod-liver oil is distributed uniformly through all of the feed. At this point the feed enriched with cod-liver oil is ready to give to the poultry. A dosage of 1 gram of cod-liver oil for 50 grams of seeds or pressed meal is recommended. One must follow definite guidelines for storing the cod-liver oil. A dark, cool room is ideal. The bottle of cod-liver oil should always be tightly closed, since the vitamin-A content quickly decomposes when exposed to air and light. For this reason, the cod-liver oil should be added daily, and the enriched food should not be allowed to stand for long, but should be fed immediately. One should make sure that the offered food is consumed immediately. The widely used cod-liver-oil emulsion contains only about 50% cod-liver oil, so that one must add 2 grams of the emulsion to 50 grams of food.

Vitamin D (calciferol) improves hatching and prevents rickets. Calciferol is particularly important in the development of the youngster, because it plays a critical role in calcium-phosphorous metabolism and participates in bone development. In adult birds, D-avitaminosis is expressed in the form of an osteomalacia, in which bone decalcification processes and a disruption of mineral metabolism occurs. Sick birds can be recognized in that they show no appetite and become emaciated. Diarrhea, drooping wings, and ruffled plumage confirm the diagnosis. In addition, the bones are soft and flexible, which is why the chickens show an uncoordinated gait and often sit on the hock joint. In anatomico-physiological experiments it was determined that the bone only contains 18% potassium salt instead of the normal 66%. Because of the decalcification, the proportion of cartilage in the bone is twice the normal amount. Accordingly, the bone becomes soft and flexible. The bill and toes are often deformed, and the chicken shows a great susceptibility for infection. At the same time, the glandular system of youngsters remains underdeveloped, so that hormonal secretion within the body is no longer fully functional and disharmony of the developmental and growth processes occurs. Frequently, D-avitaminosis ends in death. If timely countermeasures are introduced, the disease can, however, be halted and cured.

Vitamin D consists of an entire complex. One recognizes the vitamins D_1, D_2, D_3, D_4, and so forth. Vitamin D_3 is the most important, and is up to fifty times as active as vitamin D_2. In addition to the vitamins, there is also provitamin D, which is also called "sterine." Depending on whether it is formed by plants or is found in animal organisms, the biologist designates it as "phytosterine" or "zoosterine." The transformation of the sterine into vitamin D is completed through the action of the ultraviolet rays of the sun or a similar heat source.

So that calciferol can fulfill its function, the mineral salts calcium and phosphorus must be present in the correct proportion (about 2:1). If these salts are lacking, then despite the presence of a regular supply of vitamin D, rickets will develop. Additionally, a vitamin-A deficiency has a negative effect on this metabolic process. This demonstrates the complex way in which different parts of metabolic and physiological pathways interact.

For prevention and treatment one provides the birds with feed having a high vitamin-D content. Greenfood, yeast, and milk are suitable for this purpose. The most effective, however, is cod-liver oil, which contains vitamin A as well as vitamin D.

Also of importance is a supply of sunlight, since the ultraviolet rays enable the organism to convert the provitamin taken up in the food into vitamin D. Therefore, if the birds are kept indoors exclusively, lighting having an ultraviolet component is appropriate. By means of this type of light, the breeder not only prevents a D-avitaminosis, but also fosters growth, promotes laying production, and reduces rearing losses.

Based upon what has been discussed, the breeder could easily hit upon the idea of giving his birds a surfeit of vitamin D in order to protect them from the dangers of D-avitaminosis. But this would be a false step, since this easily leads to a hypervitaminosis. The consequence is diarrhea and disruption of appetite. In addition, large amounts of calcium are deposited in the blood vessels, heart, liver, kidneys, and lungs. A diminished blood supply leads to brain and heart disturbances, and the bird dies as a result of a sclerosis.

Vitamin E (tocopherol) have been designated the "fertility vitamin" because it has a positive affect on the activity and quality of the sperm, the fertility of eggs, and embryonic development. In addition, it directs the development of young birds along ordered pathways.

A typical result of vitamin-E deficiency is encephalomacia. The afflicted bird shows an uncoordinated, trembling gait and throws back its head. These symptoms of dysfunction are caused by the abnormal development of the cerebellum. Developmental disturbances often occur in the embryo, so that hatching is reduced. Not infrequently, the male fails during copulation.

Greenfood, sprouted grain, corn, soybeans, and germ oils are rich in tocopherol. Vitamin E also promotes the utilization of vitamin A in that it protects it from decomposition. Favorable results have been produced by supplying vitamin E in the form of wheat-germ oil capsules, particularly in brood cocks. Especially in the cold breeding months of December to March, this vitamin aids considerably in the production of sperm.

A vitamin F as such does not exist. Rather, it is an obsolete collective term for essential unsaturated fatty acids. Unsaturated fatty acids are essential for efficient metabolism in the organism. Additionally, they promote the fertility of the cock. Vitamin F is primarily found in the germ oils of plants.

Vitamin K (phyllochinon) regulates the activity of the circulatory system. In a vitamin-K deficiency, persistent hemorrhaging occurs. Internal bleeding particularly can have very serious consequences.

This form of avitaminosis is averted by means of abundant greenfood and carrots. To be sure, vitamin K is synthesized by the bacterial flora in the gut, but in poultry, in contrast to mammals, it is not sufficient to prevent avitaminosis, so that a vitamin-K supplement is necessary.

Water-soluble Vitamins

Vitamin B_1 (thiamine) regulates protein and carbohydrate metabolism and the water budget. The requirement for vitamin B_1 rises with increased feeding of carbohydrates (for example, soft food with a high proportion of potato).

If an underconsumption of thiamine results, at first weakness and loss of appetite appear. Subsequently, the plumage becomes rough and digestion no longer functions satisfactorily. Later, the typical symptoms of vitamin-B_1 deficiency appear. These are weakness of the extremities, unsteady coordination of movement, convulsions, and symptoms of lameness. Typical of the convulsions is the laying of the head on the back and the stretching of the legs to the front. With this outward manifestation, the appearance of death is only a question of time. If, however, only a slight deficiency occurs, then the bird merely exhibits disturbances in growth and development.

A bird that has died from a thiamine deficiency has a spongy liver, an enlarged heart, and oversize adrenal glands. As a remedy and for prevention, one feeds items containing vitamin B_1, such as brewers' yeast or sprouted seed.

Vitamin B_2 (riboflavin) affects the growth of chicks, and is stored in the organism in the liver, kidneys, and adrenal glands as well as in the blood. In the poultry it principally affects the nervous system. Within the conversion of metabolic products it promotes oxidation-reduction reactions. In this connection it should be kept in mind that an increase in protein and fat in the diet also demands a higher dose of riboflavin. If the supply of vitamin B_2 is insufficient, youngsters show loss of appetite and emaciation. As a result, development and growth come to a standstill. In the advanced stage, weakness in the legs appears; the feet slip to the side when walking, so that the birds are only able to move on the hock joints. Additionally, the toes bend out of shape and the joints swell up. Moreover, feather loss and diarrhea often result. In the slaughtered bird, the liver is fatty and degenerate. Greenfood, grain bran, sprouted seed, and brewers' yeast help to eliminate a riboflavin deficiency. Of course, the presence of vitamins B_1 and B_6 are necessary for vitamin B_2 to function. Likewise, the body needs proteins for the riboflavin to be active.

If the poultry lose feathers and show skin changes at the base of the bill, around the eyes (stuck eyelids), and on the legs, then vitamin B_6 (pyridoxine) is probably lacking. Because of a disturbed nervous system, movements become uncoordinated, and convulsions and symptoms of lameness appear. Simultaneously, laying production and body mass decrease. In general, the plumage takes on a disheveled and "loose" appearance. Physiologically, a decrease in red corpuscles occurs, since a B_6-avitaminosis stops normal blood synthesis. In addition, pyridoxine takes part in protein metabolism. Accordingly, a lack of this vitamin causes a cessation of growth. Pyridoxine is primarily contained in brewers' yeast, sprouted seeds, bran, and greenfood.

Vitamin B_{12} (cobalamin) contains the element cobalt in its molecular structural, which together with the other constituents aids embryonic development in the egg and the growth of the youngsters. In addition, it increases the utilization of vegetable proteins, so that this approaches the protein content from animal sources. For this reason, cobalamin is also called the "animal-protein factor."

As clinical symptoms of a deficiency, the birds lag behind in growth and the appetite decreases. The plumage appears in rough condition, and during embryonic development the chicks die around the seventeenth day of incubation.

In order to prevent cobalamin avitaminosis, one primarily feeds brewers' yeast, fish meal, and dairy products. If deep litter is kept in the coop, vitamin B_{12} forms on its own through bacterial transformation.

Niacin (nicotinic acid, or vitamin B_3) promotes tissue synthesis and basal metabolism; that is, it takes part in the conversion, synthesis, and degradation of proteins, carbohydrates, and fats. In addition, it has a regulating effect on digestion and blood formation.

A deficiency results in stagnation of growth and laying production. The digestive tract no longer functions normally, and symptoms of diarrhea appear. Additionally, inflammations appear in the mouth. Skin diseases occur. The legs often exhibit perosislike symptoms. The plumage has a dull and rough appearance. Hatching ability decreases, and the youngsters are prone to a delayed and interrupted feathering.

As a remedy, one feeds brewers' yeast, bran, dairy products, and carrots.

Pantothenic acid (vitamin B_5) promotes the development of young poultry and the production of adult birds. A deficiency leads to skin diseases, loss of feathers (mainly in the head and neck regions), and thickening of the foot pads. At the same time growth comes to a standstill. Dissected birds exhibit a reduced spleen and a fatty, spotted liver.

For prevention, feeding brewers' yeast, dairy products, bran, seeds, and green food is chiefly appropriate.

A deficiency in vitamin H (biotin) leads to a stagnation of growth and loss of appetite in chickens and turkeys. Simultaneously, the feathers fall out, and scaly areas of skin form. The eyelids can contract or become stuck together.

To prevent an H-avitaminosis, feed brewers' yeast, seeds, carrots, and green vegetables.

Choline (vitamin J) serves as the precursor for acetylcholine, which functions as transmitter between the nervous system and the muscles. Thus, a choline deficiency has a detrimental effect on the bird's movement. Choline promotes hatching and the development of the

Nutritional Components of Various Foods

Contents (in grams) of 100 g of Food	*Protein*	*Fat*	*Carbohydrate*	*Nutrient Ratio*
Stinging-nettle meal	12.2	3.2	28.3	1:2.9
Buttermilk	3.3	0.9	3.3	1:1.6
Fish meal	54.8	4.5	–	1:0.2
Meat-bone meal	45.6	8.5	–	1:0.4
Shrimp	51.2	2.9	–	1:0.1
Winter barley	6.4	1.1	54.6	1:8.8
Summer barley	7.9	1.4	61.0	1:8.1
Barley groats	6.3	1.4	54.7	1:9.2
Oats	7.9	3.9	41.0	1:6.3
Rolled oats	12.4	5.4	59.7	1:5.8
Oat groats	6.5	3.8	18.6	1:8.5
Potatoes	1.5	–	18.6	1:12.4
Skim milk	3.6	0.2	4.5	1:1.4
Corn	7.1	3.7	60.0	1:7.6
Corn groats	8.9	3.8	57.5	1:7.4
Rye	8.3	0.4	57.5	1:7.0
Rye bran	10.1	1.8	34.7	1:3.8
Curd cheese	26.2	0.6	0.5	1:0.07
Soy groats	37.5	1.5	23.2	1:0.7
Sunflower seed	12.1	29.0	9.6	1:6.3
Winter wheat	7.2	0.6	59.0	1:8.4
Summer wheat	8.6	0.6	57.4	1:6.8
Wheat bran	10.2	2.4	30.0	1:3.5
Wheat groats	8.0	0.9	56.4	1:7.3
Shredded sugar beets	1.4	–	74.1	1:53.0

Adapted from Baumeister, "Zusammensetzung der Futterstoffe und ihre Anwendung," *Deutscher Kleintierzüchter,* Nr. 17 (1982).

chicks as well as laying production in adults. If the laying organs are diseased, a choline deficiency is often the cause. Frequently the joints of the legs also thicken. As a result, the legs often rotate to the outside.

Choline is principally found in brewers' yeast, and is in part manufactured by the chicken or turkey itself. Doses of vitamin B_{12} reduce the choline requirement in youngsters.

Folic acid (vitamin M) plays a large role in the metabolism of youngsters primarily. Even insignificant disturbances can lead to a deficiency, which leads to a considerable decrease in production in the organism. This is expressed in poor laying and in a halt in growth and feathering. Hatching also decreases.

Folic acid, choline, and thiamine constitute a conversion system in the body. Folic acid can be provided primarily in brewers' yeast and in part in greenfood.

Vitamin C (ascorbic acid) plays an extremely important role in the organism. It has a beneficial effect on protein and carbohydrate metabolism, enzymes, and the circulation. Above all, however, it prevents infectious diseases in that it increases the body's resistance. Normally, an ascorbic-acid supplement is unnecessary, since poultry can synthesize this vitamin on their own. In situations of stress (laying period, breeding season, change of accommodations, and so forth), however, their own production often is insufficient, which leads to a vitamin-C deficiency. Disruptions of metabolism can also lead to a reduction in vitamin-C synthesis.

Greens and fruit are foods high in vitamin C. It is best to offer vitamin-C preparations to the birds in the drinking water. These are appropriate in the summer months in particular, since they help to prevent heat stress. Simultaneously, growth and condition are improved.

The so-called polyavitaminosis occurs primarily in young poultry, and is caused principally by a deficiency in vitamins A, B_1, B_2, and D. The symptoms of the disease are a swaying gait and weakness of the legs. As a consequence, the birds lie down on their sides and stretch their legs out, during which the toes usually quiver. The head is turned to the side and is placed on the back or the breast. Symptoms of diarrhea accompany this process. Death soon occurs.

To prevent a polyavitaminosis, one feeds a well-balanced diet, abundant greenfood, and a supplement of brewers' yeast. But multivitamin preparations — in the drinking water or in capsule form — also help to avert a vitamin deficiency.

In commercially prepared feed (in mash form and in pellets), but not in seeds, vitamin and mineral supplements have already been added.

Since vitamins, however, lose their effectiveness over time, the date of manufacture is always stamped on the feed packet. No more than three months should go by between the manufacture of the feed and the feeding; otherwise, the supplements in the feed lose their effectiveness in part or completely.

Hormones

Hormones are agents that bring about increased production and faster growth if used systematically. They find use in commercial poultry breeding, but are, however, of virtually no importance for hobby or purebred poultry breeding.

Antibiotics

Antibiotics retard the growth of bacteria and to some extent of viruses as well as of a number of higher organisms. Their use reduces the danger of illness produced by bacteria and has a positive effect on protein intake. With small doses of antibiotics, however, one only succeeds in eliminating weak strains of bacteria while selecting strong strains. If an organism becomes weakened, these strains can quickly produce an illness. The plants garlic, onion, and chives exhibit a natural antibiotic effect.

The Nutrient Ratio

The ratio of organic building blocks to one another must also be in balance so that optimal nutrition of the birds is ensured. In this case one speaks of a so-called nutrient ratio. According to this idea, proteins (nitrogenous compounds) on the one hand are opposed to carbohydrates and fats on the other hand. The nutrient ratio is calculated from nutritional tables, in which the digestible nutrients are broken down into fats, carbohydrates, and proteins. In the calculation, the fat-nutrient value is multiplied by a factor of 2.3 and is added to the carbohydrate content. The result is divided by the protein factor. This result finally gives the nutrient ratio. As an example, oats have 3.9% digestible fat, 41% carbohydrate, and 7.9% digestible protein. According to the above instructions, the fat is multiplied by the constant 2.3, and we obtain the value 8.97. We then add this figure to the 41% carbohydrate and obtain 49.97. We divide this value by 7.9 (protein content) and obtain a result of 6.32. Oats thus have a nutrient ratio of 6.3. This value is written as the ratio 1:6.3.

With chickens, turkeys, and guinea fowl, the nutrient ratio of the offered feed should lie between the values 1:3 and 1:9. With ducks and geese it may lie somewhat above the value 1:9. For laying chickens, the ratio of 1:3 to 1:4 has proved effective. This ratio should also be considered desirable for other kinds of poultry that are in the laying phase. With waterfowl it may be slightly higher.

As a guide for using the table, note that with a "narrow" nutrient ratio (for example, 1:4) high laying production and fast growth are attainable, and that with a "wide" nutrient ratio (for example, 1:12) growth slows down or fat deposits are forced.

Furthermore, it should be noted that the analysis values in the table only represent approximate values, since the quality of foodstuffs can vary. The table is also based on digestible nutritional values. Thus, the potato contains on average 2.3% protein, of which, however, only 1.5% is usable. For this reason the values in our table are lower than in tables that also include the indigestible nutrient values.

Nevertheless, the value of a foodstuff should not be considered in isolation but must be viewed in relation to the total diet of the poultry. Whether one should have a high or low opinion of a foodstuff depends on which range and in which dietary combination it is offered. If one considers the potato on its own, then it is a poor feed. On the other hand, it often represents an inexpensive source of carbohydrate and is readily eaten by poultry. If, for example, one combines potatoes in equal parts with skim milk and wheat bran, then one obtains a nutrient ration of 1:3.8. By means of this mixture, the potato also becomes a usable food.

With a chopping machine, the breeder can quickly shred a large amount of plant matter.

Chicken Feeding

Maintenance and Production Diets

In feeding chickens one differentiates between two principal feeding practices. These are single feeding and combination feeding. The first method is accomplished with laying-hen single feed having a raw protein content of about 16% and which can be obtained in the feed trade (feed stores). This feed is offered to one's birds all day to take in freely. No supplemental feeding of grain is provided. On the other hand, a supplemental feeding of greenfood is very desirable, since by this means numerous vitamins, minerals, and other vegetable substances are supplied to the bird. Besides the bird's health, in this way we also aid in the production of a dark-yellow to orange-colored egg yolk. Greenfood can be offered to the birds cut up or whole. In the this case, feeding in a hay-rack is advisable, since this reduces feed loss and prevents soiling the feed. For nettles, a special shredding machine has proved effective.

In combination feeding, the chicken breeder offers protein-rich laying mash (20% raw protein), and, in addition — because of the satiating and long lasting effect — between 40 and 50 grams of grain, depending on the size of the chickens, in the evening. The grain ration should consist of a mixture of wheat, corn, barley, and oats. A mixture of 30% barley, 30% oats, 25% wheat, and 15% corn has proved effective.

In winter, when a greater production of heat is necessary because of the cold, one reduces the barley portion to 20% and increases the corn ration to 25%. It has been discovered that chickens like wheat best, and then, in this order, take corn, barley, rye, and oats. Whoever does not wish to prepare the mixture himself can also buy grain ready mixed ("scratch") in the feed store.

A modified form of grain feeding is the feeding of sprouted seed, particularly the feeding of sprouted oats. In this manner one offers the chickens — particularly in winter — a supplement of greenfood. Through the sprouting process, vitamins are freed and the proportion of digestible nutrients also increases. For sprouting, one places the grain in a flat tray and keeps it constantly moist in a dark, warm place. Under these conditions it germinates rapidly. With laying hens and youngsters, the entire grain ration can be fed in sprouted form; with breeding birds, one feeds at most 15 grams per bird, since otherwise too much protein can be stored in the egg, which can prevent satisfactory hatching. If the breeder wishes to prepare the feed himself — for whatever reason — then this can be accomplished in a simple manner. He need only make sure that he adheres to the nutrient ratio and provides the essential vitamins and minerals. For example, the accompanying table shows what would be a suitable feed mixture per chicken per day (all figures in grams).

The total nutrient ratio is therefore approximately 1:3.8, which (as stated previously) is suitable for laying hens. When using meal as feed, the breeder must not forget to mix the feed components well. Whoever does not wish to go through the trouble of mixing the feed can buy the appropriate meal in mash or pellet form. In the first case the birds are kept occupied, since they must spend much time in taking in the ground feed. In this way their day passes more quickly and they hardly ever get bored, as long as the rest of the keeping conditions are suitable.

Feed Mixture per Chicken per Day
(All Values in Grams)

Feed	*Protein*	*Fat*	*Carbohydrate*
Grains:			
20 g Winter barley	1.28	0.22	10.92
10 g Winter wheat	0.72	0.06	5.90
20 g Oats	1.58	0.78	8.20
Meals:			
15 g Barley groats	0.94	0.21	8.20
10 g Oat groats	0.65	0.38	4.68
10 g Corn groats	0.89	0.38	5.75
10 g Winter bran	1.02	0.24	3.00
10 g Fish meal	5.48	0.45	–
5 g Soy groats	1.87	0.07	1.16
Total:	14.43	2.79	47.81

However, the feed lost from searching for the best particles is quite high. Filling the trough a third full has proved to be the most practical. By feeding pellets, the losses are kept within narrow limits and the birds are quickly satisfied because of the rapid feed intake. Boredom can be the consequence, from which the habit of feather picking can arise. To keep the birds busy, the feed and water troughs should be placed fairly far from each other to force the birds to walk back and forth between the two. In addition, in this way one prevents the soiling of the plumage through wet particles of feed.

For an attractive sheen of the feathers, one feeds primarily oily seeds such as hemp and sunflower seeds. Such foods are particularly suitable with black and partially black color varieties, since they bring about a green luster in the plumage. Nevertheless, this feeding must be kept within limits since the birds otherwise easily become fat.

Another significant factor that plays a role in feeding is plumage development. If one wants loose plumage — such as, for example, is typical of the Orpington breed — then soft foods are suitable, which prevents close-lying plumage. If, on the other hand, close-lying plumage is desired (for example, with Malays and Barnevelder), then the soft feed must be replaced with mash and especially with grains. By "soft feed" is meant potatoes and the like.

In summer, when the heat causes the birds discomfort, the appetite often falls off. So that the chicken still consumed enough feed, whether for satisfactory production or continuous growth, one can moisten the mash so that it is in a moist-crumbly state. In this form it is taken very readily. If the chickens begin molting, feeding must not be cut back, otherwise feather growth will be delayed and a loss in production will result. An increased amount of oily seeds can also be fed during the molt, since they give the new feathers a sheen. Also, because of their high fat content, they make available a great deal of heat for the partially featherless chickens. Corn also causes increased heat production.

For the home gardener who wishes to put his garden scraps to good use, or for the housewife who does not simply want to throw away her kitchen scraps, keeping chickens or other poultry suggests itself. All scraps should be cut up so that they are eaten easily. Bread, potatoes, noodles, rice, meat, and the like are mashed and mixed with laying mash, laying-hen complete mash, or bran and offered to the chickens in a moist-crumbly mixture. This feed is readily taken, in the same way as salad greens and the like are taken by turkeys. It is important that the leftovers do not contain strong spices, since these can have a harmful effect on the birds' health. A dressed salad, for example, is not at all suitable as chicken feed, just as spoiled food or moldy bread is not. Whoever feeds his chickens in this manner undoubtedly saves a lot of money, but the same production cannot be expected as from a properly fed chicken.

Fattening Diets

Only broiler chicks or special breeds — even though they have become rare — are suited for fattening. Such birds are bred for the purpose of putting on meat rapidly given appropriate care and diet. After about eight weeks they are ready for slaughter. Since one wants to achieve the maximum possible weight gain while providing as little feed as possible, one keeps about 15 birds per square meter. The run, otherwise obligatory, is not part of keeping for fattening. As a fattening feed, during the first ten days one offers the chicks a chick feed with a minimum of 18% raw protein, and then, until they are slaughtered, a complete feed for fattening chicks with 23–25% raw protein. Feed and water must be placed in the birds' reach within a radius of two meters, since they otherwise have to walk too far, which in my opinion is unsuitable for productive weight gain. If one wishes to prepare his own broiler feed, then protein-rich soybean meal, groats, brans, and minerals suggest themselves. If one can obtain milk products cheaply, then

Above: The birds accept moistened and stirred meal particularly readily.

Below: Planned dosage with vitamins is accomplished by forced feeding.

they can also be integrated into the fattening feed.

If a purebred-poultry breeder or a hobbyist wishes to fatten chicks of a nonfattening breed, because faulty combs, deformed toes, and the like perhaps make the birds unusable for breeding and showing, then he should realize that these chicks will never attain the weight of broiler chicks of the same age. Nevertheless, these birds exhibit faster weight gain than others of the same age with appropriate care and diet, which in this case justifies fattening. Birds fattened in this way must be slaughtered after twelve weeks at the latest, since from this time on the weight gain is no longer proportional to the food intake.

Feeding Ducks

Maintenance and Production Diets

With ducks the protein content of the feed can be lower than with chickens. All-purpose feed with 16% raw protein has proved effective for breeding ducks. For the sake of simplicity, however, one can also give them the same feed as for chickens. In the case of mash, one offers it in a more than only moist-crumbly state. In recent years, however, the feeding of pellets in a dry state has also proved its worth. Grain that may possibly be given is offered — as with chickens — in the evening. Ducks most readily take corn, then wheat, barley, oats, and rye. The feeding of sprouted grain, particularly germinated oats, has proved effective.

A proven and easily mixed laying-duck feed that can be prepared at home consists of 40 parts of cooked potato, 20 parts bran, 20 parts groats, 17 parts meat meal (or fish meal), and 3 parts of a good mineral mixture.

The offering of greenfood is also essential. The vitamin-and-mineral-rich stinging nettle has proved effective for this purpose. It is readily consumed if in a finely shredded state. Other types of greenfood should, however, also be rated positively, especially duckweed, which unfortunately can no longer be found everywhere — this should be fed only in a fresh state.

If the ducks have already laid, then one can reduce the protein content of their feed somewhat by increasing the grain ratio or by an intensified feeding of cooked and mashed potatoes. It is important that the change in diet take place gradually, since the birds will otherwise react with diminishing production and at times with a molt. The same can happen if their feed ration is suddenly increased or reduced.

Although ducks are generally considered to be the "pigs" among poultry, the same should be kept in mind with respect to the use of scraps as with chickens.

Comparative Nutritional Analysis of Commercial Poultry Feeds

Arranged in order of increasing protein and fat content. The percentage remaining (not shown in the table) may be thought of as largely carbohydrate. (All are products manufactured by Agway, Inc., of Syracuse, NY. Guaranteed analysis provided courtesy of Agway, Inc.)

Product Name	*Protein (minimum %)*	*Fat (minimum %)*	*Fiber (maximum %)*
Chickens:			
Complete Developer	12.00	3.00	5.00
Complete Grower Pellets	13.00	3.00	5.00
Shell Firm 900	14.50	2.50	4.00
High Layer 800	14.50	2.50	4.00
Complete Grower	14.50	3.50	4.00
Country Chicken Grower	14.50	3.50	4.00
Broiler Finisher	16.00	3.00	3.50
16% Cage Ration (for layers)	16.00	3.00	4.00
Nu-Cashmaker 16	16.00	3.00	4.00
Country Egg Layer	16.00	3.00	4.00
Complete Breeder 17	17.00	3.00	4.00
Nu-Cashmaker 17	17.00	3.00	4.00
Broiler Finisher	18.00	3.00	4.00
18% Cage Ration (for layers)	18.00	3.00	4.00
Pacemaker Starter	18.00	3.00	4.00
Nu-Cashmaker 18	18.00	3.00	4.00
Country Chick Starter	18.00	3.00	4.00
Complete Breeder 18	18.00	3.00	4.00
Early Lay	20.00	2.50	4.00
Country Egg Producer	21.00	3.00	4.00
Broiler Starter	21.00	3.50	4.00
Broiler Maker	21.00	3.50	4.00
Early Bird (Pre-Starter)	25.00	6.00	3.50
Gro-Mix	34.00	3.50	6.00
High Layer Concentrate	40.00	2.50	5.00
Lay Mix	42.00	3.00	5.00
Turkeys:			
Turkey Finisher	16.00	3.00	4.00
Turkey Grower	21.00	3.00	4.50
Country Turkey Grower	21.00	3.00	4.50
Turkey Starter	28.00	3.00	4.00
Country Turkey Starter	28.00	3.00	4.00
Turk-E-Pre-Starter	29.00	3.00	4.00
Turk-E-Mix	44.00	2.50	4.00
Game Birds:			
Game & Turkey Finisher	16.00	3.00	4.00
Country Game & Turkey Breeder	16.00	3.00	4.50
Country Game & Turkey Feed	16.00	3.00	8.00
Game & Turkey Grower	21.00	3.00	4.50
Gamebird Ration	27.00	2.50	5.00
Game & Turkey Starter	28.00	3.00	4.00
Shoreham Game Bird Feed	29.00	2.00	5.00

Above: Malays attain the imposing height of 80 centimeters. **Below:** Belgian Games are one of the largest breeds in the game group.

Facing page: Since chicks feed themselves immediately upon hatching, it is easy to provide them with appropriate food.

Fattening Diets

Today one begins fattening ducks from the first day on. Here too, specialized fattening duck breeds are preferable to others. If the ducks are already three to four weeks old, then one can still begin fattening them. With ducklings, one feeds prepared duckling feed with a protein content of 18%. After 14 days one switches to fattening-duck complete feed. At the same time the ducks are placed in a small stall and a small enclosure. Because of the limited room for movement they gain weight very rapidly. If the ducks show an aversion to prepared fattening feed they can also be fattened with potatoes, kitchen scraps, and groats. If milk products are also mixed with the feed, the protein content will be increased. These also improve the meat quality. The latter method of feeding has proved to be particularly effective with adult ducks that are to be fattened for a short time before they are slaughtered. Vitamin-and-mineral-rich greenfood is indispensable for fattening ducks. If one provides the ducks with supplemental light in the coop, they can also feed during the night, which leads to faster and greater weight gain. With laying ducks the light also leads to increased production, given appropriate production feed. As a rule, they should be given about 14 to 16 hours of light.

Feeding Geese

Maintenance Diet

Since geese are basically grazing animals, a large meadow run is appropriate for them. As a rule, the goose-breeding hobbyist cannot offer his birds one of these. As a compromise he must feed his geese large quantities of greenfood. Moreover, he also saves on expensive concentrated feed in this way. Among greenfoods, stinging nettle — as with other kinds of poultry — again predominates. Through its nitrogen-rich compounds it is very beneficial to the goose's production and growth.

In addition, a so-called self-prepared maintenance feed which prevents the geese from becoming too fat can be used. A suitable feed, for example, is a soft feed composed of equal parts of groats, shredded carrots, and potatoes. A feeding of grain in the evening has proved effective; wheat should, however, not be included. The grain ration should be about 75 grams. Since the geese very readily take oats, one can feed them from time to time in sprouted form. An additional maintenance feed consists of 130 grams of oats, 100 grams of potatoes, 50 grams of coarse barley, and 25 grams of shredded carrots. With a simple diet, however, one can certainly also give the geese laying-hen complete feed or breeding-duck complete feed, which is offered in a moist-crumbly form with some softened bread or potatoes. As a rule, feeding is done in the morning and evening.

Fattening Diets

If one wishes to fatten a Christmas goose, then it should not be given much room for exercise so that the desired weight is put on rapidly. A proverb states that "rest is half of fattening"; that is to say, unnecessary disturbance of the birds prevents rapid weight gain. After the rearing period, from the fourth or fifth week on, the young geese receive corn, barley, oats, cooked potatoes, softened bread, kitchen scraps, and greenfood. It is important to keep in mind that the birds' diet must not be changed overnight, but that it must be switched gradually by continuously shifting the ratio from the rearing feed to the fattening feed. Another simple fattening feed mixture would be a groats and potato diet, offered in moist-crumbly form. One can feed as much as the birds will eat. Since geese also feed during the night, one should give them twice the grain ration at night as with maintenance feeding. Several days before the slaughtering date one should start to feed only pure grain. Twenty-four hours before, they are no longer fed at all so that the digestive tract is empty. Changing the drinking water frequently is also important. Stale and soiled water has a negative effect on the goose's appetite and health.

For fattening young geese (fattening from the first day on), the following feed mixture has proved effective, which one feeds four times a day during the first four weeks: 12% wheat bran, 30% barley groats, 15% corn groats, 20% oat groats, 10% protein concentrate, 5% dry yeast, 5% soybean groats, and 3% minerals (subsequently, switch over to fattening feed). The geese are also fed with mixed greenfoods. Of course, one should only use suitable breeds for fattening, since only they make use of the fattening feed most efficiently and then exhibit the desired rapid growth. Additionally, feed

Buying feed in bulk is the most cost-effective course.

Above: If one feeds meal or pellets, the feed dispensers must be filled regularly.

Facing page: A rooster of the Mediterranean type.

companies manufacture goose-fattening mixtures.

Turkey Feeding

Maintenance Diets

Turkeys require a large run, since they otherwise do not thrive even with the best diet. There they find a large part of their nourishment themselves; otherwise one gives adult turkeys the same feed as chickens. Regular feeding of vitamin supplements is important, since their requirement is considerably higher than that of chickens. If the essential vitamin supply is lacking, turkeys can quickly become ill and die. The feeding of copious amounts of greenfood is advisable with them as well. A special complete feed for breeding turkeys, which contains 16% raw protein, is available in the feed trade. One must make sure that the expiration date for vitamins and other supplements has not already expired. Otherwise, one should avoid the purchase of feed of this kind. Meal feed should be offered in the form of pellets. Turkeys accept mash feed particularly readily in a moist-crumbly state. If one wishes to prepare one's own turkey mash, the following recipe can be used: 5% fish meal, 5% meat meal, 10% soybean meal, 15% alfalfa meal, 15% wheat bran, 40.24% grain meal, 1% bone meal, 2% calcium meal, 1% iodized salt, 0.01% manganese sulfate, 5% dried whey, 0.3% food oil with vitamins A and D, 0.15% vitamin-D supplement, 0.3% riboflavin. Because of the difficulty in obtaining and mixing this feed, it is best to depend on prepared feeds. A small supplemental feeding of grain in the evening has a positive effect.

Fattening Diets

For fattening, during the first six weeks one feeds complete feed for turkey poults with 28% raw protein; during the seventh to twelfth weeks, complete feed for fattening turkeys type A with 24% raw protein; during the thirteenth to eighteenth weeks, complete feed for fattening turkeys type B with about 20% raw protein; and during the nineteenth to twenty-second weeks, complete feed for fattening turkeys Type C with 15% raw protein. Home-made turkey-poult fattening feed consists of 20% barley groats, 20% corn groats, 20% wheat groats, 10% wheat bran, 20% protein concentrate, 3% feed yeast, 5% mineral mixture, and 2% hormone concentrate. In addition, a medicine for the dreaded blackhead disease should be fed (this has already been added to the complete feeds).

Feeding Guinea Fowl

The feeding of guinea fowl does not differ from that of chickens. A laying-hen complete feed with a raw protein content of about 16% is satisfactory. With the supplemental feeding of grain, one should — as with chickens — feed the more protein-rich laying mash. The guinea fowl's feed requirement is about 100 grams a day.

Drinking Water

As a general rule, drinking water is critical for all poultry species. Decreases in production are often associated with an insufficient supply of drinking water. The reasons for this deficit are mostly because the water is too cold in winter or too lukewarm and therefore stale in summer. For optimal metabolism the birds need water that is fresh daily and the correct temperature. Microorganisms, which reduce water quality, quickly multiply in stagnant water. The presence of algae, bacteria, and the like not only produce poor egg flavor and an unfavorable embryonic development in the egg, but also decrease resistance to disease and the stability of the body. Often a bird becomes infected with an illness through such water. During the summer, one should place the water in an appropriate drinking container in a shaded location so that it stays fresh for as long as possible. For the winter, heated troughs or heatable bases are required; these ensure water of the proper temperature while using little electricity. Each time before the troughs are filled, they should be cleaned thoroughly.

To keep ducks in a manner that is suitable for them, they should be offered an area of water. With dry keeping at least a water basin must be available. Since ducklings are very sensitive to wet ground, their drinking dish should be placed on a special wire grating with a container to catch any water that is spilled. Regular cleaning of the watering places is one of the most urgent duties of the duck breeder, since otherwise the coop easily becomes damp or the run will quickly become soiled. When keeping ducks in a pond, one must pay attention to the highest possible water quality, since otherwise illnesses can easily become a threat.

Geese are given a water container which they cannot enter. A separate bathing opportunity is most appropriate.

In addition, drinking water is suitable for the administration of vitamin preparations or for drinking-water inoculations. By "drinking-water inoculations" is meant administering small viral doses that stimulate the bird's body to produce antibodies.

Above: A foot chain inhibits the excessive fighting drive found in some young roosters.

Facing page: Bantams—Silkies above, a mixed flock below.

GENETICS AND BREEDING

The Reasons for Breeding

In order to maintain the quality of a breed or to even improve upon it, we must perform the task of breeding by mating chosen brood fowl (assembling the breeding stock) and subsequent selection of offspring characteristic of the breed. For this purpose there are various procedures in poultry breeding with which one can reach the desired goal. Nevertheless, before one can engage in planned breeding for particular characteristics of an appropriate breed, one must familiarize one's self with the basic laws of inheritance. These will be introduced briefly in the following.

Chromosomes — The Carriers of the Genes

In every cell is found a nucleus which directs all of the metabolic processes in the cell and thence in the body. In the nucleus are found — well-packed in protein envelopes — the genetic factors, or genes. Inside of the protein envelopes they are arranged in long, twisted double chains. Each double chain with its protein envelope is called a "chromosome." Chromosomes are the carriers of the genes, which regulate the life processes in the body and also transfer the specific characteristics of a breed to the offspring in the course of reproduction. Every cell nucleus of a bird species has a particular number of chromosomes, and the chromosome pairs themselves particular shapes. Different species have different chromosome numbers and shapes, which is the basis for preventing inbreeding (except for occasional infertile hybrids) between species. If a crossing resulting in fertile offspring nevertheless succeeds (for example, between Red and Grey junglefowl), this means that the species are closely related to each other.

Inside the chromosomes we find the genes, which are responsible for the expression of external characteristics, and also for a properly functioning metabolism, organ growth, and for much more. A gene does not directly cause the expression of the character but only the synthesis of a chemical substance that usually causes the character to be expressed through its action on many metabolic pathways. Frequently, many genes, which have complex interdependence among one another, are required to accomplish this. If a gene in such a system becomes inactive, then the expression of this character can be prevented, which sometimes also leads to the death of the embryo in the egg.

For the sake of simplicity, however, we will proceed from the assumption that one gene is directly responsible for one character, since one can then plan the breeding much more readily. In the "back of one's mind" one should always be aware that the matter is much more complicated than is shown in the breeding diagrams presented, and that these are based on a simplification of the biological processes.

Homozygosity and Heterozygosity

In all body cells, with the exception of the sex cells, one always finds two paired chromosomes, both of which always possess a gene for the same character in the same place. One of these chromosomes comes from the mother, the other from the father. If the genes at both sites are factors for a particular trait (for example, for red eye color in poultry), then one speaks of "homozygosity" with respect to this character. If the gene on the paternal chromosome expresses red eye color, and the maternal gene expresses yellow, then it is a question of "heterozygosity."

Diagram: Homozygosity and Heterozygosity

Genes for red (x) and yellow (o) eye color

Homozygosity: x x

Heterozygosity: x o

Father's Mother's

Chromosomes

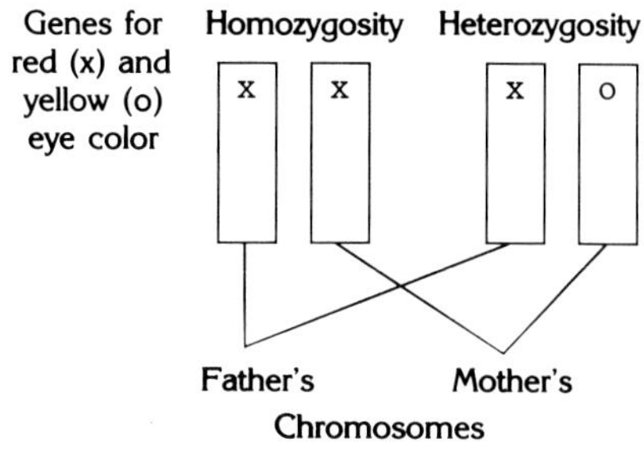

Diagram: Meiosis with Homozygous Genes

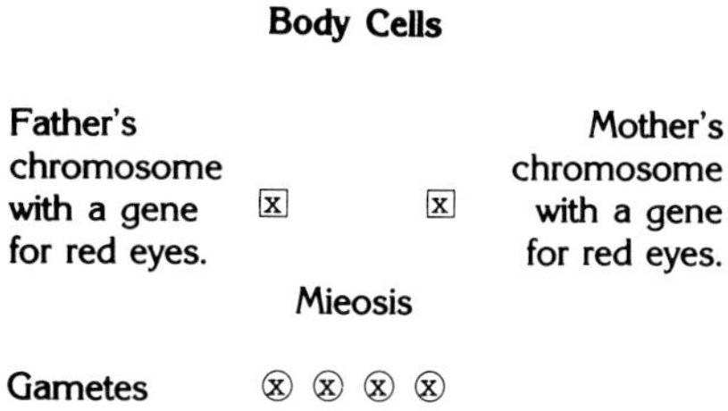

Body Cells

Father's chromosome with a gene for red eyes. [x]

[x] Mother's chromosome with a gene for red eyes.

Mieosis

Gametes ⓧ ⓧ ⓧ ⓧ

Diagram: Meisis with Heterozygous Genes

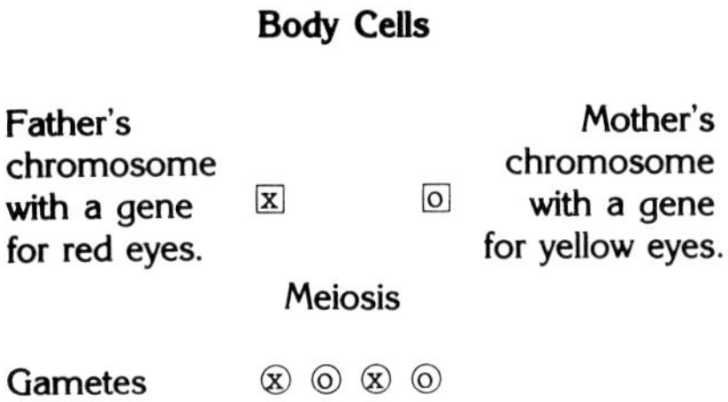

Body Cells

Father's chromosome with a gene for red eyes. [x]

[o] Mother's chromosome with a gene for yellow eyes.

Meiosis

Gametes ⓧ ⓞ ⓧ ⓞ

The Swan Goose of Asia plays a part in the ancestry of some breeds of domestic geese.

Above: A mixed flock of ducks, probably none of which are purebred.

Below: Piebaldness in geese produces markings which vary greatly from individual to individual.

In reproduction the chromosomes present in pairs are separated and are always transported into different reproductive cells (gametes). Each gamete therefore has only one of the two original chromosomes. This takes place with all chromosome pairs in the cell nucleus. Whether the original maternal or the paternal chromosome reaches any particular gamete is determined by chance. If the cells were homozygous, then this question has no significance for inheritance. The situation is different with heterozygosity. In this case, different genes are transferred during the formation of gametes and in reproduction. The process of the division of the chromosome pairs with the corresponding genes to the gametes is called "meiosis" (reductive-division). The following diagram should clarify this important process.

After meiosis all the gametes have half of the chromosome number. When two gametes fuse with each other during fertilization, they form a body cell, which represents the starting point for the new life. The fusion of the paternal and maternal gametes can result in either homozygous or heterozygous cells.

Depending on the results of the combination of the paternal and maternal genes in the individual cells, one speaks of "intermediate" or "dominant-recessive" modes of inheritance.

Intermediate Inheritance

Color inheritance represents an instructive example of intermediate inheritance. If the mother is homozygous white and the father is homozygous black, then the offspring are gray (example: Andalusians, whose gray-blue coloration is nonetheless designated as "blue" by breeders), since they possess the gene for white as well as one for black, and both genes combine in the expression of color. Hence, in the mating of the offspring with one another white, gray, and black offspring can be produced, namely in the ratio 1:2:1 (whereby the given ratio is dependent on a very high number of offspring).

Dominant-Recessive Inheritance

In dominant-recessive inheritance, the gene for black, for example, is expressed more strongly (dominant) than is the gene for white (recessive). Although all of the offspring possess the genes for black and white as alleles, they are uniformly black in the expression of their color, since, of course, the gene for black masks the gene for white. In this connection one speaks of a uniform "phenotype" (expression of external characters) and a nonuniform "genotype" (genetic makeup). Since the gene for white is

Diagram: Intermediate Inheritance

(Adapted from Engelmann 1975, *Vererbungsgrundlagen und Zuchtmethoden beim Geflügel)*

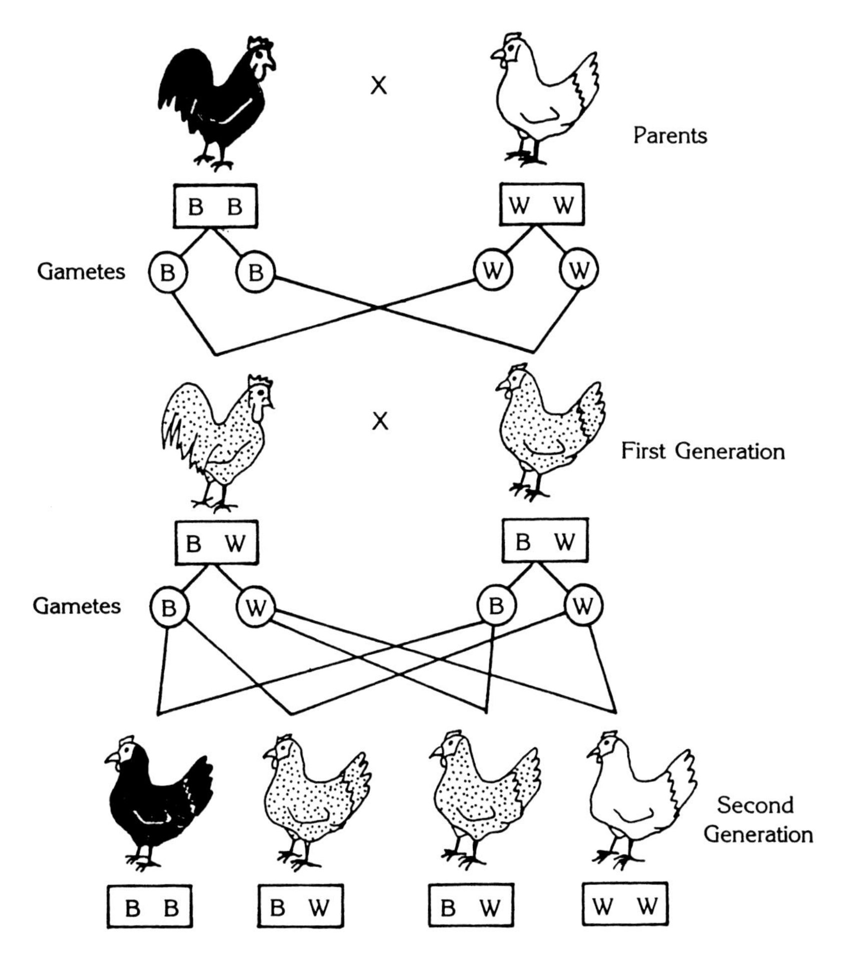

B = Factor for black plumage
W = Factor for white plumage
BB = Homozygous for black
WW = Homozygous for white
BW = Split for white and black, which is expressed as gray plumage

Above: In banding four-toed breeds, the back toe is pressed against the leg.

Below: The typical way of handling a chicken: the index finger is located between the legs, which are held firmly by the thumb and the middle finger.

Above: By using a lattice to collect the eggs, the chicks can be assigned to their mother with certainty.

Below: An Old English Game brood hen cares for her young lovingly.

Diagram: Dominant-Recessive Inheritance

(Adapted from Engelmann 1975, *Vererbungsgrundlagen und Zuchtmethoden beim Geflügel)*

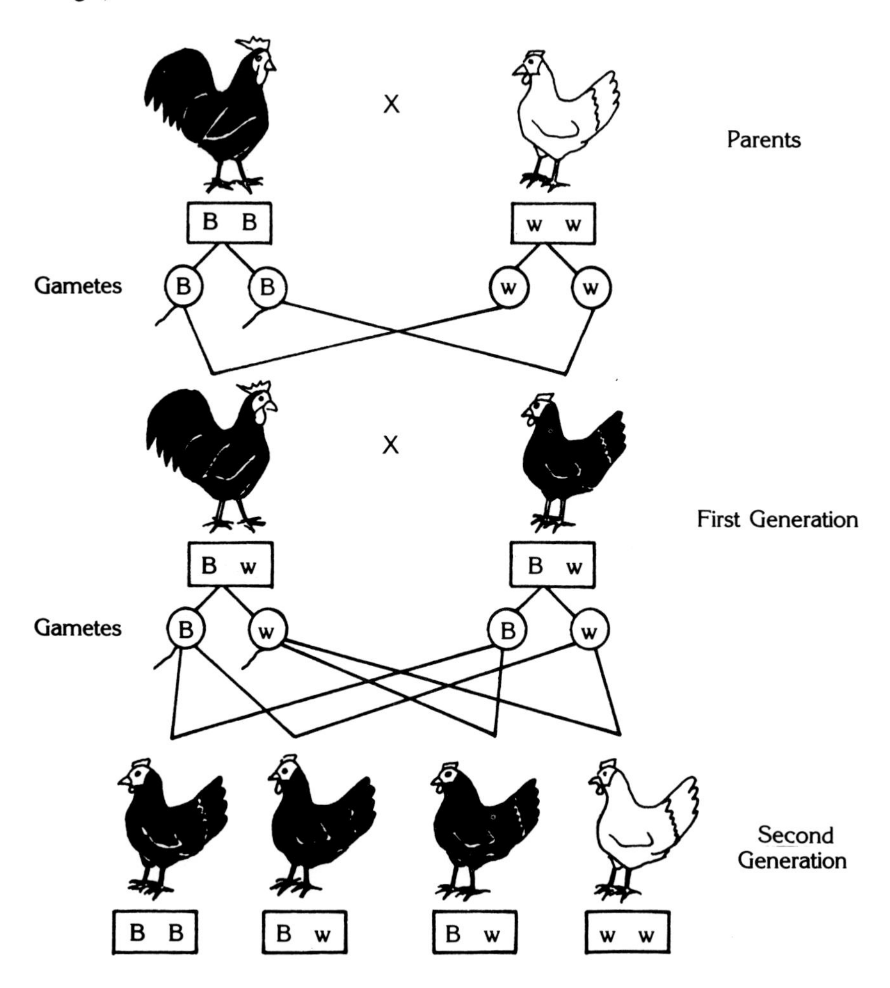

B = Dominant factor for black plumage
w = Recessive factor for white plumage
BB = Homozygous black (dominant)
ww = Homozygous white (recessive)
Bw = Genotypically split for black and white, phenotypically black because of the dominant factor

present in the genetic makeup, in a crossing of chickens of the first generation, homozygous black, heterozygous black, and homozygous white birds again occur in the ratio 1:2:1. This result relates to the genotype. If one only considers the phenotype, however, then one arrives at the ratio 3:1, since heterozygous and homozygous black birds look identical (phenotypically).

Problems of Inheritance

Both inheritance modes are relatively simple, since the expression of the character is limited to a single gene. Often, however, several genes on different chromosomes are responsible for the expression of one character, or the expression of one character is linked with another. The latter case occurs principally in sex-linked inheritance. For these and other reasons inheritance cannot always be explained by means of the intermediate and dominant-recessive modes just described. During the reductive-division (meiosis) the process of "crossing over" can also occur. In crossing over homologous portions of the chromosomes are exchanged. Spontaneous "mutations" (the sudden change in one or more genes) also occur. Several kinds of mutations are recognized: deletions, in which portions of the chromosomes break off and do not reattach; duplications, in which a part of a chromosome is doubled; and translocations, in which a part of one chromosome breaks off and attaches to another chromosome. These two processes, crossing over and mutation, together with subsequent selection — whether by nature itself or induced by man — are certainly the basis for continued evolution in the case of the individual organism and therefore of species and breeds, whereby the latter can also approach an ideal goal for man's purposes through crossing with other breeds.

It is essential for the breeder to know that genes are inherited with a certain range of variation, that is, a character has a certain degree of play in its phenotypic expression within which it can unfold. Through suitable accommodations and attentive care this degree of play can to some extent be shifted in a favorable direction. One then speaks of a "modifying influence." Despite this fact, one cannot make a very good bird from a genetically poor specimen. But one can certainly foster the traits of a good bird, and in this way, for example, achieve better results at a poultry show.

Breeding Schemes

In regard to breeding schemes we recognize mainly inbreeding, modified inbreeding in the form of line breeding, and crossbreeding.

Above: The Rouen duck is very well suited for fattening.

Below: The Muscovy drake is larger than the duck and more heavily carunculated.

Above: Clean the water container each time before refilling it.

Below: If a bird has external parasites, it should be treated with a powdered insecticide formulated for poultry.

Inbreeding

In inbreeding, one mates the parents with the children or the children among one another. In this manner one achieves a certain degree of homozygosity if one continually selects for particular traits. Through regular backcrosses with a parent or a closely related bird (for example, a son with the mother, a nephew with the grandmother) and subsequent planned selection, we can fix one or more traits of the initial bird. We achieve a similar result through the mating of siblings. Here too a basic condition is the selection for phenotypic traits, and, when possible, also for genotype with the aid of inheritance diagrams. Through years of continued inbreeding we obtain a breeding stock which is homozygous with respect to the desired trait. In this way we can produce high-quality birds without large breeding quotas.

Crossbreeding

If one practices inbreeding for many years — with the goal of homozygosity — then "reductions" (reversions) with respect to vitality, hatching success, production, disease resistance, and the like, can appear. In this connection one speaks of "inbreeding defects." For this reason one is forced to cross in unrelated birds of the same breed (occasionally also of another breed) — this is crossbreeding. Through crossing, we introduce new genetic material into our breeding stock, which again leads to an increase in production and a renewal of vigor. Nevertheless we should not wait until inbreeding defects appear but should introduce an unrelated bird into the breeding stock early on. This is best accomplished with a female bird. Although we examine the phenotypic traits of such an bird as closely as possible for its suitability for crossing into our breeding stock before the purchase, it is quite possible that it will not meet our expectations genotypically. In the case of a female only this bird's chicks will be "duds." If on the other hand we cross in a male bird with a defective genotype, then all of the offspring will carry the undesired genetic traits, and it will take years of work before we again reach the point we had already achieved before the crossing. If unrelated birds are crossed into the poultry breeding flock, then, as a rule of thumb, it can take five years or more before inbreeding defects appear.

Line Breeding

Line breeding is a combination of inbreeding and crossbreeding. Line breeding is accomplished by building up two separate inbreeding lines using two unrelated birds which, however, correspond to each other in the characteristics of the breed. In both lines one carries out suitable backcrosses and sibling matings. In this manner one maintains two separate inbred lines for years, in which one attains homozygosity in as many of the desired characteristics as possible. By removing one of these birds from the line and mating it with a foreign bird one can build up another line. After several years of separate line breeding, one can then take a female bird from one of the lines and cross it with a male bird of the other line. In this way one prevents possible inbreeding defects without having to cross in a bird from unrelated foreign stock. The result is that the genetic material of the stock is, in a manner of speaking, kept at the same point. The direct offspring of the crossing of two lines often exhibit a surprisingly great increase in production. One speaks here of a "luxuriating effect." This effect also appears with normal crossbreeding. With line breeding it is important that one constantly build up new lines that are not or are only slightly related to one another and are pure in relation to the other line.

The starting birds for line breeding need not necessarily correspond in all of their traits. Nevertheless, the weaknesses of one mate must be balanced by traits in the other that are typical of the breed. Through rigorous selection in the subsequent generations, one then obtains a homozygosity that is typical of the breed with respect to the characteristic that was poor in one of the initial birds. If one mate is supposed to balance out the weakness of the other through breeding, we speak of a "balancing mating."

Chicken Breeding

Brood Stock

The breeding flock represents the basis for high-quality breeding. As with every kind of poultry, chickens must at least harmonize with one another phenotypically when breeding begins, and later, with insight into the inheritance modalities, one can also harmonize the breeding stock genotypically (see the discussions of inbreeding and line breeding) with one another. A breeder often needs years, however, to determine the specific genetic traits of a breed. For the purebred-poultry breeder who takes birds to exhibitions, it should be noted that the most expensive birds at shows are not necessarily the best birds for breeding. When purchasing breeding stock, one should not place to much emphasis on the purchase price, since with the purchased breeding birds one lays the cornerstone for the future quality of the birds. If the birds are not of high quality initially, then the breeder is forced to obtain quality through quantity and subsequent selection. The cost of raising a large flock of chicks exceeds the purchase price of a high-quality breeding stock many times over; one must be satis-

Above: A grating of wire mesh and a removable tray in the brooder makes for hygienic chick rearing. **Below:** With a round automatic feeder, the right amount of food is always available, and the young birds have sufficient room to feed.

Facing page: More than other duck breeds, the Muscovy is inclined to spend time on land rather than in the water.

fied with a small number of offspring, but one does obtain quality birds.

Number of Brood Birds

If a breeder already owns a laying flock, he must keep his breeding flock separate from it. The breeding flock consists of three to five hens and a brood cock that is compatible with the female birds. In general, one can estimate up to twenty hens for a breeding flock of a light breed, and with heavy breeds up to ten hens; but with such a large number of birds control is lost, and one usually does not have many birds that match one another genetically. As a rule, problems of fertility also occur. The rooster's ejaculate is often of little or no value after several matings, so that the hens mated last lay infertile eggs. These problems do not crop up with a small breeding flock; each mated hen receives high-quality ejaculate.

Social Hierarchy in the Breeding Flock

Generally, the hens at the top of the pecking order are uncooperative and often refuse to mate for a fairly long time. In a large breeding flock a hen of this kind is hardly ever mated, and its offspring are correspondingly few. In contrast, hens lower on the pecking order almost always allow themselves to be mated. For this reason, breeders with large breeding flocks run the risk of obtaining offspring primarily from hens with less ability to hold their own in the social hierarchy. The usually unavoidable consequence is a decrease in the vigor and condition of the offspring. In small breeding flocks even the hens at the top of the pecking order are mated at regular intervals. The brood cock should have been one of the highest-ranked birds in regard to vigor in the rooster flock, which is kept separately during rearing. He guarantees vigor, condition, and vitality for the following generation. If a hen regularly lays infertile eggs, then one can assume that the rooster finds her "unsympathetic," and therefore does not mate with her. If one integrates the hen — if possible — into another breeding flock, she may then lay perfectly fertile eggs.

Designing a Brood Pen

So that the breeding birds can get used to one another early, the flock should be put together as soon as possible. After the competition for positions in the social hierarchy, the birds become acclimated to the new environment of the breeding coop and accept it as their territory. A coop design that produces blind corners should be rated highly, since in this manner lower-ranked hens can withdraw from the view of higher-ranked hens, and thus can live in a less stressful environment. Higher production and improved egg quality are the reward for dividing the coop into various individual domains. The run should be laid out in exactly the same way; the birds should not, however, be allowed in the run in winter, since cold weather has a negative effect on egg production. On springlike and especially on sunny days, on the other hand, staying outdoors is very beneficial, since sunshine has an extremely favorable effect on the chickens' metabolism.

The coop should have good ventilation, although drafts should not be allowed to develop. The litter must be dry. A sand bath is also indispensable for chickens in the coop, since it contributes considerably to the birds' hygiene and contentment. Since dust bathing is an inborn instinct that must be exercised, the presence of a bath of this kind is psychologically relevant and ultimately also affects the birds' physiology. The addition of an insecticide powder for ectoparasites to the sand bath aids in the fight against poultry parasites.

Trap Nests and Inspection

The nests should be placed in sheltered, relatively dark areas of the coop. To be able to ascertain parentage, these should have a trap-nest mechanism; that is to say, after the hen enters the nest, the trap-nest mechanism prevents it from leaving again or a different hen from entering the nest. By inspecting the trap nests several times a day, the breeder knows exactly which hen laid which egg. We mark the eggs with the band number or some other sign, so that after the chicks hatch in the hatching compartment of the incubator those of each hen can be identified. (The methods of marking will be discussed in their proper place.) Without this procedure, planned breeding is hardly possible, since only in this way can we discover the genotypic composition. In this fashion, after several years of line breeding a breeder can predict which brood cock and which hen produce the best offspring.

The procedure described is often not possible for the working person, which is why the spouse is usually "allowed" to take over the responsibility of inspection. As an alternative to the certainly time-consuming procedure described, the breeder can limit himself to a trap-nest inspection on weekends. In this case he also knows precisely which egg comes from which hen. Assuming the eggs can mostly be easily distinguished from one another by shape, shell formation, size, and color intensity, one will soon be in the position to assign the eggs laid to the appropriate hen even without a nest inspection. This method of course works only with a small breeding flock. With a large flock, the complexity factor overpowers even the best-intentioned efforts.

Above: The pied Muscovy has a particularly appealing appearance.

Below: The difference in size between the massive Rouen and the dainty Call ducks is patent.

Effects of Temperature and Light

The most favorable temperature for chicken keeping is between 15 and 19° C. In order to maintain such temperatures — especially in winter — we must invest some money. Since the birds continue to lay satisfactorily at low temperatures, we can in part neglect the matter of heating. Nevertheless, the temperature should not fall below 5° C., if only because below this temperature the hatching eggs lose their ability to develop.

More important than the coop temperature is artificially lengthening the day, for the internal rhythm of life and therefore laying production is influenced by the external factor of light. The eye collects the light and converts it into a nervous impulse. This impulse travels through the midbrain and reaches the hypothalamus (pituitary gland), which it then activates. As a result the hypothalamus produces so-called secondary sex hormones, which — depending on the sex — affect either the male or female gonads. Moreover, this mechanism only functions upon exposure to a specific period of light. In winter the period of daylight is too short to trigger this biologically directed process. Through "dispensing" artificial light, which is easily controlled with a timer, the short winter day is lengthened and the production of the secondary sex hormones is stimulated. Since the rooster requires a longer period of time for the production of sperm of optimum quality than the hen for egg formation, it is desirable that the cock be given light three weeks before the hens. Depending on the breed, the hens need about two to three weeks after the light is supplied before they begin to produce eggs. Breeders of heavy breeds in particular must therefore start their preparations for breeding early. With these breeders the lights often are already burning in the coops in November or December, and in January the first chicks are hatched. The photoperiod should be about 14 hours a day.

Feeding the Breeding Flock

In addition to these technical factors involved in fowl care, the diet of the breeding birds plays a decisive role. As with the diet for laying hens, a mash diet of laying-hen complete mash with 16 to 17% raw protein has proved to be beneficial. It ensures that too much protein does not collect in the egg, which cannot be completely reabsorbed (taken up) and then leads to the death of the ready-to-hatch chicks. Besides the ration of laying mash, a supplemental feeding of grain of 50 grams per bird per day should be included. The food value of the grain lowers the high protein intake from the laying mash somewhat, which is advantageous during the breeding phase. The laying mash must be easily accessible to the breeding hens, and the feed container must be of the right size. The often-extolled feeding of sprouted grain must be approached with a certain degree of caution during the breeding phase, since this stimulates an enormous increase in protein in the brood egg. The feeding of soft feed should be avoided with the breeding flock, but the occasional supplemental feeding of greenfood is beneficial. Greenfood strongly supports the vitamin supply of the breeding chickens, and indirectly supports the developing ability of the hatching eggs. Since feeding greens presents considerable difficulty in winter especially, the breeder is forced to administer vitamins preparations. Liquid vitamin solutions, soluble vitamin tablets, and vitamin capsules are available for this purpose. The first two preparations are given in the drinking water, while the vitamin capsules must be force-fed to the birds. Multivitamin preparations, which contain all vitamins in the correct formulation, are advantageous. Vitamin E can be fed as a supplement to the brood cock rooster in the form of wheat-germ oil capsules, since the so-called fertility vitamin markedly improves the formation and quality of sperm. Carrots have a high vitamin-A content, which is why they are often fed in winter as a substitute for greenfood.

Worming

The matter of diet also involves worming the breeding flock. Pharmaceutical companies offer numerous preparations for this purpose. When using them the directions should be followed exactly. Feeding pieces of onion and garlic also provide a certain degree of prevention. But despite this measure, worming during the breeding season should not be neglected. Afterwards, vitamins help the body, which was put under stress by the worming, to quickly recuperate.

Duck Breeding

Breeding Stock

For ducks, a run with bathing facilities is very important; it strongly supports the fertility of the brood eggs. The ideal situation would have a run area of 15 square meters per bird; this can, however, only be realized in the rarest cases. As a rule the ducks should be provided with a grass run or an area of pure sand. If one furnishes the run and the bathing basin like a natural environment with respect to vegetation, then this will be to the psychological benefit of the breeding stock. Getting the birds used to one another early is also very important psychologically, since ducks present certain problems in this regard.

Coop, Nesting, and Space Requirements

In the duck coop especially, the ventilation must work well, since ducks release most of the water they take in through respiration in the form of water vapor. Ducks likewise require dry litter. After being mated by the drake, who lives together with his three to five ducks, the breeding ducks begin with egg laying. Frequently they do not lay their eggs in the nests (nest boxes with straw and hay) provided for that purpose, rather they lay or bury them in the coop litter or in the run, which is why one must be careful when searching for the eggs. By placing artificial eggs in the prepared nests, one can induce some ducks to lay their eggs there. The so-called reed huts of ornamental waterfowl have also proved effective. For these one ties 1½ meter lengths of reed together into bundles as big as one's fist, and joins these together into a round straw hut that tapers to a point on top, having a diameter of 25 to 40 centimeters at the base, depending on the duck breed. Space for a small entrance should be left in the hut. One can place it in the coop or outdoors; as a rule, this nest is accepted very readily. It is important that the floor have a small depression, which is lined with hay, reed leaves, or reed seed heads. If one wishes the ducks to lay exclusively in the coop, then one may not let them out of doors before ten o'clock; they will have almost always laid by this time. Here too, a trap nest, which is 30 centimeters wide, 45 centimeters deep, and 60 centimeters high, is of course ideal. If the ducks do not enter the trap nest on their own, then one can put them in there in the evening, so that they lay their eggs there in the morning. Since all ducks lay in the morning, one needs as many trap nests as one has breeding ducks. With heavy ducks one can allow about eight months for full development. Therefore one must begin breeding correspondingly early, forcing production and the onset of laying. In this respect artificial lighting is absolutely necessary. Light breeds do not require artificial lengthening of the day, and will begin with egg production from about the middle of March on.

It is important to prevent any alarming of the ducks. This is especially true during darkness. Strange people can also contribute to the birds becoming nervous. Once the duck flock has lost its confidingness, it is often difficult to tame them again. In the daily contact with the birds, hurried and sudden movements should be avoided under all conditions, since the birds could react with a decrease in production.

Feeding the Breeding Flock

The ducks' diet consists of mash and grain feed during the breeding season. Mash feed is made up of the following components: 10% bran; 70% groats; 15% animal protein (for example, fish or meat meal); 2% dry yeast (feed yeast); 3% mineral mixture for poultry.

Alternatively, ready-mixed duck-breeding feed or laying-hen complete mash can be fed. It is given in a moist-crumbly form in the morning and the evening. Often the birds take it more readily when the consistency of the mixed feed is slightly watery-to-pasty. Feeding potatoes should be avoided with duck breeding stock. To supplement the mash diet the ducks receive a grain mixture, in which the grain — especially oats — can be germinated. Shredded carrots are a substitute for greenfood in winter; in the spring one reverts to young green plants, particularly stinging nettle.

Goose Breeding

Breeding Stock

Geese, though they are in a real sense grazing animals, are waterfowl. For this reason bathing facilities are necessary for them, and not only for fertility. Not later than November must one decide which geese one wishes to use for breeding. The breeding geese must be placed together in December, so that they will have gotten accustomed to one other before the breeding season. This acclimation is an essential prerequisite for good fertility, particularly if new geese are added to the breeding stock.

It is advantageous to only use adult birds for breeding, since the goslings, or youngsters, as a rule are not as vigorous and fast growing. One should allot four to five geese to a gander. One can breed a gander for about three years and a goose up to ten.

Breeding Space

The coop is the place where the breeding goose stays, sleeps, broods, and raises young. The goose coop as well must have good ventilation. Dry litter is not only essential for production, but for health as well.

Even in cold weather one should let the breeding geese out of doors and provide access to water. Because of their dense down layer, they are very well protected against the cold. Fresh air has a beneficial effect on their constitution. In severe cold or with snow it is better, however, to leave them in the coop or to allow them only a short stay out of doors.

Nesting

For early egg production, one should provide them — as with chickens — with artificial light. Once the goose starts laying, the first egg is marked with a pencil. One leaves it in the nest, where it should prevent the subsequent eggs from being misplaced. For hatching the first egg is worthless, since it is usually infertile. All of the eggs that are then laid are taken from the nest and kept in a cool place (8 to at most 15° C). It is important to mark the eggs with the date laid and the mother. To compensate for the removed eggs one can place artificial eggs in the nest. When the goose stops laying and lines its nest with down, then the incubation phase has begun. Now one returns the eggs to the mother, with the oldest given one day earlier than the most recent. Often the geese do not continue to incubate, so artificial incubation may be necessary.

Feeding the Breeding Flock

For the breeding season one changes gradually from maintenance feed to laying feed. Laying-hen complete feed or breeding-duck feed, which is fed in a moist-crumbly state, has proved effective. In addition, the breeder offers 40 grams of sprouted and 30 grams of unsprouted oats. The supplementary feeding of greens or alternately the feeding of carrots in winter is essential. Giving cod-liver oil with the grain provides the geese with vitamins A and D in concentrated form.

If one wishes to prepare the feed mixture himself, he includes 15% bran; 65% groats; 5% fish meal; 10% protein concentrate; 2% yeast (feed yeast); 3% mineral mixture for poultry.

Laying-duck mixture can also be used alternatively as mixed feed for geese. One feeds about 120 grams of it daily — divided between the morning and the evening hours. In addition, coarse grit must be freely available to the birds.

Turkey Breeding

Breeding Stock

With turkeys one allows up to twenty hens for each tom with the lighter breeds, while breeding toms of heavier breeds should have a maximum of fifteen hens. As with chickens, however, a much smaller breeding flock is recommended, since each bird requires about one square meter of coop area. An extensive run, which allows the birds sufficient room to move, is important. Hedges should be used to provide shelter from cold ground winds, since these have an extremely negative effect on the fertility rate.

Coop and Nests

The nests consist of a sturdy laying box lined with straw and hay. Since the poults' development requires up to eight months, one must begin breeding correspondingly early; artificial lighting also stimulates egg production in this case. To prevent diseases, the coop must be dry, and the air circulation should be optimal.

Feeding the Breeding Flock

For feeding one uses turkey breeding feed, which is obtainable ready-mixed in the trade. One can also use laying-hen complete feed, so long as breeder does not forget the necessary supplemental vitamin-and-mineral supply. One should plan on 200 grams of mash feed per bird per day. A supplemental feeding of 75 grams of grain per bird per day is necessary, as is a supply of greenfood. A feed mixture that can be prepared at home consists of 17% corn groats; 15% wheat meal; 15% wheat bran; 15% oat groats; 5% alfalfa meal; 10% meat meal; 5% dried milk; 15% soybean meal; 2% mussel shells; 1% salt.

Because of its high protein content, a supplemental feeding of grain is essential.

Guinea Fowl Breeding

Breeding Stock

One should allow three to four hens for each cock. As a rule, he will not accept any more, since he originally was monogamous during the breeding season. If one wants youngsters early, then in this case too lengthening the day with artificial lighting suggests itself. However, the chicks must then be reared artificially, since they need a great deal of warmth. For natural rearing one should postpone breeding until May, at which time no supplemental lighting is needed.

Coop and Nests

Since guinea fowl, which on the whole receive the same care as chickens, have a tendency to hide their eggs, one should place their nests in a slightly elevated position in dark areas of the coop. The eggs are taken from the nest daily (replacing them with artificial eggs). For breeding, two-year-old hens should be given preference to one-year-olds because of their better constitution.

Feeding the Breeding Flock

Guinea fowl receive the same feed as chickens. But a feed somewhat higher in protein content is also usable. Here too supplying minerals and vitamins as well as the feeding of greenfood is important.

THE EGG

The Egg in the Human Diet

The egg is of great importance in the human diet, since in its protein, fat, and carbohydrate components it contains substances that are channeled into growth and energy production in the body. The egg provides vitamins, minerals, and other substances that regulate metabolic processes. These processes do not involve just the building of body tissue but also its degradation and conversion. For example, half of our muscle tissue is replaced in approximately 160 days. Even the liver renews half of itself in ten days. For these processes our bodies require protein as a building block. Depending on its composition, the minimum protein requirement is about 30 to 40 grams a day. The optimal amount is about twice that. As a guiding principle, the following serves for adults: one gram of protein per kilogram of body weight per day. Proteins — including those found in eggs — are made up of twenty different building blocks, the so-called amino acids. The human body has the ability to manufacture only twelve of them. The remaining eight must be obtained from the diet, because the human being cannot live without them. It is important, however, not only that the amino acids be present in the food; the proportions in which they occur is also significant. We then speak of either the high or low "value" of the protein.

In general, the ratio of the different amino acids in animal protein is better suited to the human constitution than is vegetable protein, though both types of protein can complement each other to reach an optimum result. The egg's biological value should be rated very highly in this regard, and at the same time it complements the value of other nutrients. A comparison of foods will serve to clarify this. The value of the egg is about 94%, milk and curd cheese "only" have 86%, and potatoes, a source of vegetable protein, only show a percentage of 67.

In addition to protein, the egg also supplies the body with fat. For the ideal utilization of fat, not only is its composition of importance, but so is its ratio to protein. The fat-protein ratio is about 2:1 in the egg, which is nearly optimal for the human body. As a rule, the fat component of other foods is higher.

Concerning carbohydrates, it should be noted that they constitute a negligible portion of the chicken egg. On the other hand, vitamins are very plentiful in the egg. In general, eggs belong to the most vitamin-rich foods of all. Especially the vitamins A, B_1, B_2, D, and K are present in large amounts.

Of the essential minerals, the egg contains principally calcium, phosphorus, potassium, sodium, and chlorine. Other significant components of the chicken egg are lecithin and cholesterol. Lecithin makes up 9% of the egg. The egg is thus the only food that contains lecithin in large enough amounts for the human organism. It contributes to proper functioning of the brain and nerves. Lecithin remains present in the egg for months without appearing to decompose. Cholesterol, on the other hand, of which as much as 280 milligrams is present in an egg, is feared by many people. Cholesterol is the precursor for the hormones of the adrenal gland, gallic acid, and the vitamins A and E. The human body manufactures most of its own cholesterol; only 10% is obtained from in the diet. The dreaded arteriosclerosis develops when, among other substances, cholesterol and calcium are deposited on the arterial walls. This disease is, however, promoted less by cholesterol intake than by smoking, lack of exercise, and obesity. Possibly even a hereditary factor, an inherited susceptibility to this disease, plays a role.

Quality of the Fresh Egg

It is important to the user that eggs be fresh. This is often very difficult for the layman to determine. The size of the air space offers a simple means. The larger it is, the more moisture the egg has lost and therefore the older it is. The air space can be seen by testing (candling) it with a strong light. Alternatively, one can also place the egg in water. If it is fresh, it will lie horizontally on the bottom of the container. The older the egg is, the more vertically it stands in the water. An old egg even rises to the surface. So that eggs stay fresh as long as possible they are refrigerated. According to the German consumer standard for eggs of 1975, only such eggs that were stored over 8° C. and that have an air space of no more than 6 millimeters may be designated as "fresh" (Grade A). Additional items complete the requirements for the fresh egg. Eggs that were stored under 8° C., have an air chamber of no more than 9 millimeters, and exhibit a few other particular characteristics, traits, are considered to have been "stored" (Grade B). However, one must distinguish between administrative regulations and biological facts.

With storage at 0° C., at a relative humidity of 75%, eggs can be made to last for up to a year. Through refrigeration the chemical and biological reactions are reduced. Simultaneously, the growth conditions for harmful bacteria, which are widely known to grow best in moist, warm conditions, are changed for the worse.

In selling eggs the following weight classes are significant: Class 1: more than 70 grams; Class 2: 65 to 70 grams; Class 3: 60 to 65 grams; Class 4: 55 to 60 grams; Class 5: 50 to 55 grams; Class 6: 45 to 50 grams; Class 7: under 45 grams.

The egg carton bears various inscriptions. Besides the weight class and the grade, the week number is registered on the carton. In the particular week listed on the carton, the eggs were graded into their weight class and packed. They were usually laid earlier. Some egg cartons carry the designation "extra" on a label. This means that the eggs are especially fresh. This label must be removed no later than a week after packing, since at this point in time the special category "extra" no longer exists.

Composition of the Egg

A chicken egg consists of 11.6% shell and 88.4% contents. The contents of the egg are made up of approximately 73.7% water, 12.6% protein, 12% fat, 0.7% of nonnitrogenous substances, and 1.1% minerals.

The egg white is not pure protein, but instead consists of 86.7% water, 11.2% nitrogenous substances (protein), 0.5% fat, 0.9% carbohydrate, and 0.5% minerals.

The yolk is made up of the following substances: 49% water, 21.6% fat, 9.1% lecithin, 15.7% protein, 0.4% cholesterol, 0.3% cerebrin, 3.3% minerals, and 0.6% of other substances.

Of the kinds of protein present in the yolk, the ferruginous and phosphorated vitellin, or ovovitellin, is the best known. This is apparently not a matter of a pure protein, but rather of a protein-lecithin complex, since one can obtain up to 30% lecithin from vitellin.

For the purebred-poultry breeder and hobbyist who sells surplus eggs on the side, the yolk is of great interest, since no doubt virtually every German housewife is delighted when she cracks open an egg and a dark-orange yolk is revealed. One then often hears the argument that this is a "biological egg" or, as it were, an egg from happy chickens. Such eggs increase the demand on the producer.

But what is really going on with the coloration of the yolk? The coloration is caused by pigments that belong to the fat-soluble carotenoids and to the xanthophyll pigments. The pigments cannot be manufactured by the bird itself; they must therefore be supplied through the diet. Such pigments occur in all green parts of plants. When a leaf colors in the fall, these are visible to everyone. But they are also contained in corn, carrots, and in many other foods. In the body the pigments enter the body fat, the legs, the beak, and of course also the yolk. The feeding of foods poor in pigments, such as potatoes or rice, leads to pale yolks. Besides feeding greens, one can also intensify the color of the yolk by providing small doses of pigment vehicles (0.35 milligram capsanthin per 100 grams of feed).

It is interesting that the egg yolks of hens that lay frequently are less intensely colored than those from hens that lay few eggs. The reason for this is that the hens that lay intensively have a greater pigment requirement than the poorer-laying hens.

Egg yolk contains two different kinds of yolk: white and yellow yolk. The white (lower in fat) yolk surrounds the ball of yolk like a cloak. In some spots the more intensely colored yellow yolk is sometimes visible. One then speaks of "yellow" or "marbled" eggs.

Formation of the Egg

The yolk with its egg cell (blastodisc) has its origin in the ovary and at first only grows very slowly, but then quickens its growth greatly. In the ovary, which in birds is not paired as it is in mammals, the membrane enclosing the ripe yolk tears open and releases the yolk (yolk with the egg cell) from the ovary. It reaches the oviduct with the aid of cilia. There, for about fifteen minutes, it capable of being fertilized by one of the rooster's sperm cells. Subsequently it encloses itself in a thin membrane (yolk membrane). If no sperm cell has fertilized the egg cell thus far, the egg remains unfertilized. If fertilization did take place, then embryonic development already begins. During the egg yolk's passage through the oviduct, first viscous, then watery, and finally again viscous egg white is deposited around the yolk.

The secretion of the egg whites occurs through the mucous membranes of the wall of the oviduct. Since the oviduct's mucous membrane exhibits spiral folds in cross-section, on its journey the egg moves through the oviduct spirally, so that the viscous egg white on both poles of the egg are twisted into the albuminous threads familiar to everyone, which are also known as chalazae. Their function is to stabilize the yolk in the middle of the egg. They may possibly also offer the remaining egg-white structure an anchoring point.

In the subsequently narrower part of the oviduct, a shell membrane, which is made up of two layers, is deposited around the viscous egg white. Newly formed watery egg white is passed through this membrane. This egg white is also produced by the oviduct's mucous membrane. Finally the oviduct's shell glands secrete the calcium shell. The glands obtain the calcium via the circulatory system. The calcium for egg formation is not assimilated until two to three days before the formation of the egg shell. In contrast, in the contents of the egg, elements from feedings of 40 days previous are present. Calcium is deposited directly on the outermost shell membrane, whereby a protein binds both together. After about seventeen hours, the 0.3-centimeter-thick shell is finished. The calcium deposition takes place in small, adjoining columns, so that small intermediate spaces remain between the tiny columns, which narrow toward the top side of the egg. Over the calcium layer is located the so-called cuticle (mucous layer, or elastic epidermis), which exhibits a smooth surface structure because of fatty proteins. The cuticle is applied by cells of the uterus. (In the uterus are found the so-called sperm-storage glands for fertilization without immediate copulation.) The cuticle sometimes contains — depending on the breed — color pigments that give the white shell a brown (for example, Barnevelder or Welsumer) or a green (Araucana) color. With Araucanas, through a unique mechanism of these chickens, not only the cuticle but the entire calcium shell is colored green. If one cracks open a brown and a green egg and examines the shell from the inside, then it will be white in the former and greenish in the latter. The pigment of the brown and green eggs is derived from the breakdown of blood corpuscles. Therefore, since the brown color is only deposited externally on the calcium shell, the argument of many users that "brown eggs taste better than white ones" is based more by psychological than biological grounds. To be sure, there are differences in flavor between eggs from battery-kept chickens and from chickens raised out of doors with fresh feed. The free-range chickens have a significantly better and distinctive flavor.

In addition, it has been publicized in the meantime that the eggs from chicken farms are often the "final resting place" for medications. The use of drugs is necessary in large-scale poultry farming, so that diseases can be nipped in the bud. Otherwise the disease could spread in an epidemic and destroy the commercial basis of the egg producers. In the more humane conditions of keeping of chickens in large flocks, more medications are used than in the crueler battery situation. The large-flock egg is therefore anything but valuable.

Abnormalities of the Egg

Besides the typical chicken egg, a number of curious types of eggs also exist. For example, if two yolk balls ripen simultaneously in the ovary and pass through the oviduct together, then both of them will be incorporated into one egg, making the amazing double-yolked eggs. So-called yolk-free eggs are just the opposite. In these eggs, the egg-white glands of the oviduct are not stimulated to produce egg white by the egg yolks, but rather by small foreign bodies (for example, blood clots). Consequently, the egg remains without a yolk. If an egg yolk which is released by the ovary is not taken up by the oviduct cilia and reaches the abdominal cavity, then it will swell and mature there. If this process is repeated several times, then one speaks of "layered eggs." If these layered eggs (abdominal-cavity eggs) rot, they can endanger the hen's life. Abdominal-cavity eggs often occur when the hen is very old and the oviduct has consequently lost tone. Feeding excessive protein, over-breeding, and overexertion can also cause the formation of layered eggs under certain circumstances. Eggs with blood spots are a biologically completely harmless variation. A blood spot can, for example, come into the egg when a capillary ruptures during the egg yolk's passage through the oviduct. Such blood spots are not detrimental to the flavor of the egg but do perhaps give it an unappetizing appearance. Eggs without shells often occur with young hens that are just beginning to lay. These are unobjectionable with youngsters in that phase, since egg laying has not yet "come into full swing." If shell-less eggs occur frequently, however, then this points to an insufficient supply of calcium. For this reason one should always make mussel calcium or other calcium preparation available freely to one's birds, so that the metabolism can work with an optimal calcium supply. If an egg remains in the oviduct too long or the calcium glands overfunction, then eggs can receive an abnormal deposit of calcium.

From time to time one also hears or reads about so-called fertile eggs. By these are meant chicken eggs that have been incubated for nine days. If one eats them, they are supposed to help combat numerous illnesses. Whether these actually have a health value remains to be seen; nevertheless, in our culture they are not a normal food from an aesthetic point of view. Of the duck egg, which, incidentally, is produced differently from the chicken egg, one often hears that it tastes unappetizing and transmits diseases. This objection is justified when the ducks drink from polluted or fetid bodies of water and lay their eggs in the dropping-contaminated soil of untended runs. There the eggs can become infected with bacteria or fungus. But feeding spoiled foods and the intake of less-than-perfect water can have at least a negative effect

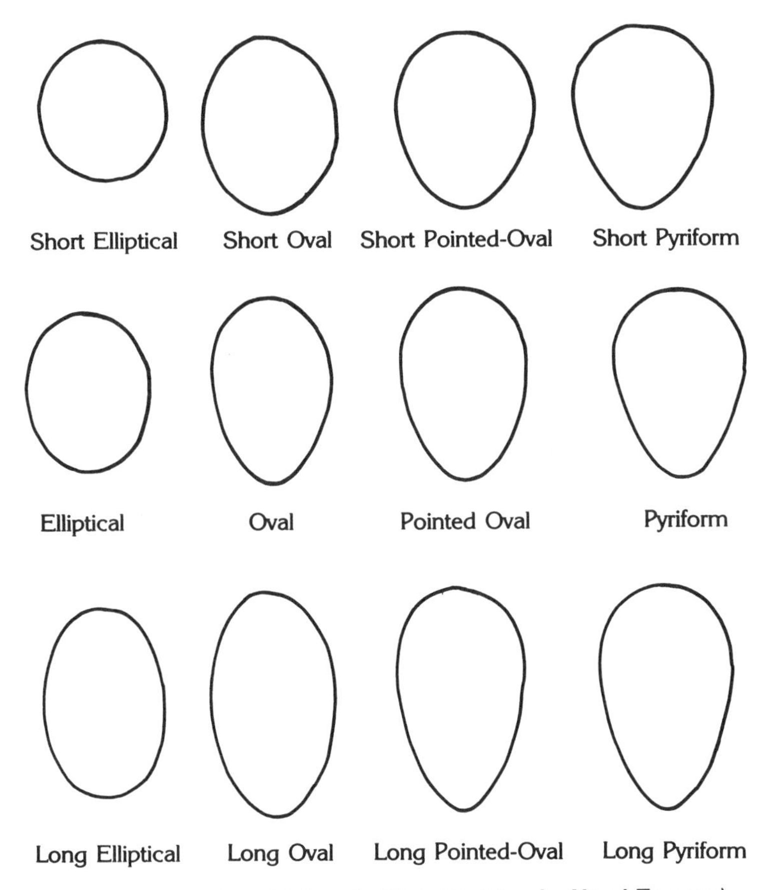

Egg Shapes (Taken from Makatsch 1974, *Die Eier der Vögel Europas)*

on the flavor of the egg. Likewise, excessive feeding of fish meal — as with all poultry — can also give an unpleasant flavor. If ducks are given clean, fresh water and nutritious feed, and are kept and cared for in a regular fashion — including daily removal of eggs — then the eggs will be perfectly satisfactory (to the extent that the bird does not itself carry an infection). Ducks eggs are usually even more delicious than chicken eggs. They also have a more nutritious yolk than chicken eggs, and, as a rule, weigh more.

Egg Shape

With respect to shape, one can distinguish four basic types: elliptical, oval, pointed oval, and pyriform eggs. From these basic types, different variations, such as short or long oval eggs or long and short pyriform eggs are produced. The oval, and perhaps the pointed oval, may be considered the "classic" egg shape. At purebred-poultry shows one also sees eggs from time to time. There the oval shape is also considered to be the ideal.

The Hatching Egg

During the breeding season, how the hatching eggs are handled is of great importance for breeders. Basically, we remove the eggs from the laying nest daily and mark them with the laying date and parentage. For writing we always use a pencil. The eggs are stored in a place where the temperature lies between 8° C. and a maximum of 15° C. If the temperature drops below 5° C. the eggs lose their ability to develop, or malformations are produced. In addition, the storage place should be sunless and free of drafts, and should not contain dry air (75% relative humidity is ideal). Additionally, the area should be free of odors,

Hatching eggs are stored on the pointed end or on their side; in the latter instance, they must be turned.

Eggs with an uneven surface are not suitable for hatching.

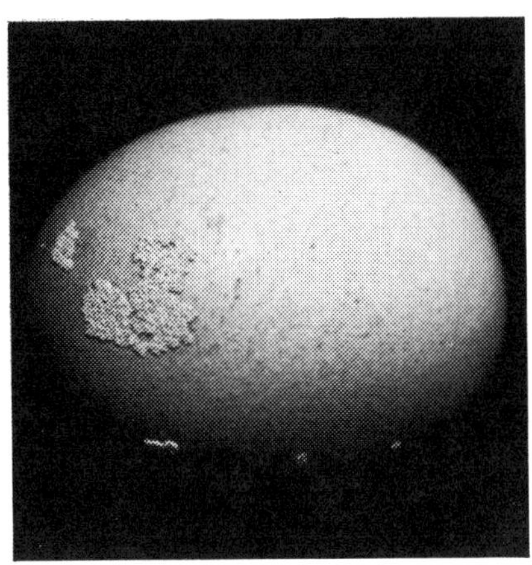

Eggs with patchy calcium deposits are not hatchable.

since the eggs could pick them up.

The eggs are stored in an egg carton, standing on the pointed end. Alternately, they can be stored lying on their sides, but then they must be rotated half way around their long axis at least daily, so that the yolk does not sink through the egg white layers and stick to the egg shell or to the shell membrane during storage.

Nothing that is universally valid can be stated concerning the age of the hatching eggs. In general, the eggs should be as fresh as possible. However, even four-week-old eggs still show satisfactory hatching results. If both older and fresh eggs are mixed in the clutch, the older eggs should be incubated somewhat earlier (about half a day to a whole day), so that they can make up the developmental deficit they have relative to fresh eggs, which will allow them to hatch together. The cleanliness of the eggs is very important for hatching. The pores of the shells must not be clogged with dirt. It would be incorrect to scrub dirty eggs, since the pores of the eggs would be blocked even more and the protective cuticle removed. As a result, microorganisms could easily enter the egg. Poor hatching would be the inevitable consequence. With soiled eggs, the particles of dirt should be scratched off with a knife, or the eggs should be placed in lukewarm water and the dirt removed without rubbing.

Eggs with a faulty air space are unsuitable for incubating. The air space must lie at the blunt end of the egg, between the shell membranes. When viewing the egg with a strong light one can see it very clearly.

Also unsuitable for hatching are eggs that have a foreign body (for example, a blood spot) or a crack in the shell. The cracks are often so fine that one can see them scarcely or not at all. Only with candling does the defective shell become apparent. During incubation the egg white, which becomes yellowish as it congeals, often emerges from the cracked areas. Also unsuitable are eggs with calcium rings and additional deposits of calcium. They break easily since their calcium structure is defective. Their moisture regulation does not function precisely either, so that a failure — particularly with artificial incubation — is unavoidable. In addition, the calcium shells are often porous at the places with deposits, and the yellowish egg white will emerge in small droplets during the first days of incubation.

Additionally, hatching eggs must be the correct weight (the so-called minimum hatching weight). With eggs that are too large there is the possibility that they have two yolks, or that the chick will not be able to completely assimilate all of the egg white and consequently will not hatch properly. As a rule, small chicks hatch from eggs that are much too small. The danger also exists that the tendency to lay small eggs will be inherited. The correct hatching-egg minimum weight can be found in the standards promulgated by poultry associations. In general, the minimum weight with heavy- and medium-weight breeds lies between 53 and 60 grams, and with light breeds between 48 and 55 grams. Bantam eggs weigh about 40 grams. It should be obvious that only eggs with the classic shape should be used for hatching. Distorted, indented, elliptical, and similarly deformed eggs are useless.

If one has been sent the hatching eggs, they must be allowed to set quietly for 48 hours before they are placed under the hen or in the incubator. Otherwise the shocks associated with transport will have a negative effect on the hatching results.

BREEDING AND REARING

Poultry Development

Depending on whether one breeds heavy, light, or bantam breeds, one must begin hatching correspondingly early. One must also consider that roosters require about one to one and-a-half months longer to come into condition than females. Accordingly, heavy breeds are hatched in January-February, medium weight breeds in February-March, light breeds in March-April, and bantams in March-April-May. The higher the outdoor temperatures are, the better the chicks will develop. With ducks we can wait until they begin laying on their own (in March), but the use of electric lighting — above all with heavy breeds — is certainly feasible. Since ducks grow rapidly, their development by fall is ensured. At 200 days, as a rule, they can be considered full-grown.

Geese reach "adulthood" at 270 to 300 days. With heavy breeds, supplemental lighting is therefore justifiable.

With turkeys, growth varies, depending on the color variety. The birds are sexually mature by about the thirty-second week of life. Heavy birds should be bred earlier than light ones. Artificial lighting is necessary at times.

In order to obtain early guinea fowl chicks, we must "dispense" electric light to our breeding stock.

For the purebred-poultry breeder who shows his birds, the times given do not apply if he wishes to enter his birds at exhibitions early in the year. In this case, he must calculate the developmental period back from the time of the show in order to arrive at the ideal hatching time. Accordingly, the time of egg collection and the acclimatization of the breeding flock, as well as the effects of technical breeding factors (for example, providing light) must be calculated. One must also note that the conditions in summer have a more positive effect on development than the conditions in fall and winter.

If the breeder has collected his hatching eggs, then two alternatives offer themselves for incubating: natural incubation by a broody hen or artificial incubation in an incubator. The breeder usually turns to the incubator, since a broody hen is rarely available at the time she is needed. In addition, the brooding instinct has been bred out of most breeds, so that the birds rarely attempt to brood at all. Nevertheless, we do still have a few relatively dependable brooding chickens, such as the Araucana and the Silkie Bantam. However, from time to time we also obtain a broody hen in other breeds.

The incubation period is the same with natural or artificial incubation, and amounts to:

Chickens	21 days
Bantams	21 days
Ducks	27–29 days
Muscovies	35 days
Geese	29–30 days
Turkeys	27–28 days
Guinea Fowl	26 days

If we are not restricted by the hatching date, we can safely give a brooding hen several eggs; seldom is there a more beautiful scene than the leading and care of the chicks by the broody hen. The touching sight of the mother-and-children bird family finally provides us a few hours of contemplation and recovery from our stressful everyday life, and has especially great psychological significance in allowing mostly alienated city-dwellers to experience nature.

Natural Incubation

The Broody Hen

NEST

One undertakes natural incubation with chickens primarily, which is why this will be discussed first. The chicken coop is an ideal hatching area for broody hens, since they can remain in their accustomed surroundings and brood on the nest of their own choice. The nest must be protected against the entry of other laying chickens, since rivalries can otherwise result in damage to the clutch. With a trap nest one simply lowers the flap and the broody hen has her separate brooding place. A piece of material that partially covers the opening provides the broody hen with the preferred semidarkness.

If brooding in the coop is not possible, then a semi-dark room with sufficient humidity and without a draft (for example, a cellar) is suitable for brooding. Disturbances should be avoided in any case.

When preparing the nest, one should disinfect of parasites with a powder or liquid. To line the nest we use moist-crumbly garden soil or sections of sod, which we place upside-down in the nest. In this substrate we form a depression. For nesting material we add either hay or short straw.

The moistness of the soil promotes embryonic development in the egg. Since this moisture will evaporate after a time, we must moisten the soil regularly.

It has also been shown, however, that a simple hay or straw nest produces good hatching results.

Brood hen incubating.

Care and Keeping

Before the broody hen is set in the nest, we apply a suitable insecticide powder to her plumage, setting the clutch in the nest. A heavy hen can cover about 16 eggs, a light hen only 12. Basically, the broody hen should be able to cover the eggs completely. Evening is the best time to set the broody hen. Usually she goes into the nest on her own. Otherwise one lifts her up carefully and makes sure that her feet are placed between the eggs. Especially in the beginning is the soothing semidarkness of the nest important. When brooding in a chicken coop, the broody hen should only be given grain in the evening, in that one carefully lifts her from the nest and lets her pick along with the other chickens. When taking the hen from the nest, it is best to hold her with one hand placed under her body and then to lift her off the nest. Outside she will defecate, take in grain and water, and bathe very readily in the dust bath.

If the broody hen is kept separately, we also take her from the nest in the described manner and allow her to take in feed and water from a dish we have provided. Here too provision for a sand bath is appropriate. If incubation takes place in a separate room where the broody hen has the opportunity to leave the nest, then we give her feed and fresh water ad libitum at a distance of about one meter from the nest. The broody hen will then provide for herself.

After about seven days, and again after about seventeen days, we candle the eggs with a strong light and remove the infertile and dead eggs.

Three days before hatching, that is, on the seventeenth day, we moisten the soil in the bottom of the nest once more, after candling the eggs, and then leave the hatching to the broody hen. On the nineteenth day we take her from the nest once more, and subsequently avoid any further disturbance. On the twentieth day of incubation the chicks pip the shell with the egg tooth and hatch at most 14 hours later — on the twenty-first day. Before we bring the broody hen into a separate rearing area, we leave her — if possible — on the nest for 24 hours. At that point the chicks have dried and are ready for the "daily routine."

With confinement in a small cage, a hen quickly loses her desire to sit.

The Brooding Duck

Domestic ducks (with the exception of domestic Mallards and Muscovies) as a rule incubate poorly or not at all, so we incubate duck eggs artificially or under a chicken or turkey hen.

The Brooding Goose

In contrast, geese to some extent sit very reliably. For hatching, one provides each goose with her own nest box, which has a size of about 60 to 80 centimeters, in a dark corner of the coop. The nest is equipped with a framework which prevents the goose from leaving the nest alone and without the breeder's notice. By this means the breeder prevents a mix-up of the nests. The eggs collected during the laying phase are returned to the goose as soon as she has lined her nest with down and has stopped laying. The older eggs are placed under her one day before the freshest. The clutch of a goose should consist of at most twelve eggs. After ten days the eggs are candled with a strong light, and the infertile ones are removed from the nest. If one started artificial incubation at the same time, the breeder can put fertile eggs from the incubator in the nest as a substitute for the infertile ones. So that the eggs are provided with sufficient humidity, it is beneficial to also make a bath available to the goose during the incubation period. Otherwise, bathing the eggs — particularly in the final days — in water with a temperature of 38° C. has proved effective (the same applies for duck eggs). Also very beneficial is a moistening of the soil that constitutes the bottom layer of the nest. Hatching takes place after about 30 days, at times somewhat earlier or later. One must not disturb the goose during hatching, since otherwise she could trample the goose chicks (goslings) underfoot out of sheer excitement.

The Turkey Broody Hen

The Nest and the Possibility of Forced Brooding

Turkeys brood reliably, and usually begin to do so on their own. One can also carry out forced brooding with them. This is successful, however, only with birds that have not yet layed. A place in a semidark coop is the most favorable site for forced brooding. A wreath woven of hard straw and lined with hay serves as the nest site. The nest substrate is formed in the same way as with the chicken broody hen. The turkey is set on the nest and covered with a wicker basket, which prevents it from getting up. It is best to weigh down the basket, since at first the turkey usually attempts to leave the unfamiliar surroundings. Warmed artificial eggs will stimulate the brooding drive. After one day we remove the turkey and clean the nest while she feeds and drinks. Subsequently she is again placed on the nest. After about 24–28 hours the turkey hen usually will have accepted her brooding duties; at this point the hatching brood eggs are placed under her. In this fashion one can let turkey hens brood two to three times in succession.

One hen receives 24 chicken eggs, 18 turkey or duck eggs, or 30 guinea fowl eggs.

Forced Brooding with the Chicken Broody Hen

We can also carry out the same forced brooding, with variable success, using chicken hens of medium-weight and heavy breeds. However, the hen becomes "broody" only after four or five days.

Breaking the Brooding Habit

At this juncture the reverse of brooding — discouraging brooding — should not be overlooked. If a breeder uses artificial incubation entirely, it is exasperating to him when a breeding hen from which he wants as many hatching eggs as possible becomes broody. In this case one places the broody hen in a homemade, ventilated brooding-disaccustoming cage. This cage has a rectangular shape; its edges are constructed of wooden laths, the sides and the floor of wire mesh. This cage hangs free in the room or is hung on a wall; it is equipped with feed and water containers. If the hen is locked inside, she will stop brooding after at most 6 days, and will lay again after an additional 14 days.

Alternatively, one can also place the hen in a litter-free coop with a rooster. As a result, the hen will lose her broodiness quite quickly.

Duck nests attached to the wire of the enclosure.

Artificial Incubation

For good hatching results with artificial incubation, appropriate keeping and feeding of the breeding stock are as necessary as optimal hatching conditions. Artificial incubation is based on so-called hatching factors. Only when these are exactly right do the eggs have optimal hatching conditions. Many breeds show somewhat variable demands with respect to the hatching parameters, so that the breeder often must experiment a bit at first. In general, however, at the start of his experience in technical hatching he should follow the directions provided by the manufacturer of the incubator.

Hatching Factors

By "hatching factors" the breeder understands the matters of temperature, humidity, turning the eggs, and oxygen supply. Hatching factors are somewhat variable with each poultry species. The accompanying table provides a generalized outline.

Temperature

A continual exchange of heat takes place between the egg and the environment. For this reason the incubation temperature is doubtless the most important factor in the hatching process. Since the egg constantly gives off heat (the heat is created as a result of metabolic activity in the egg), differences in temperature result. So that the egg temperature stays at a constant level, it is necessary to regulate the temperature. This is accomplished by means of the thermostatic regulation of the incubator and constant air exchange. In chickens, 37.8 to 38.0° C. should be considered a standard value. If temperature deviations from the given standard value occur over a fairly long period of time, a decrease in hatching can occur. An average deviation of 0.2° C., however, is not enough to cause negative results. A decrease in temperature of about 1° C. from the standard value causes a delay in the hatching date and unsatisfactory vitality of the chicks. If the temperature falls by about 2° C., embryonic development will be slowed and poor hatching will result. At a temperature of less than 34° C., hatching does not occur at all. Elevated temperatures have a more serious effect on the second half of the incubation period and are as a whole more harmful than lowered temperatures. Slightly elevated temperatures accelerate the hatching date. The optimal value often differs with the various breeds. Araucana chickens, for example, hatch better at a constant incubation temperature of 38.3° C. than at 37.8° C. With too high a temperature, the accelerated

With a homemade candling device, one can determine the fertility of hatching eggs after about seven days.

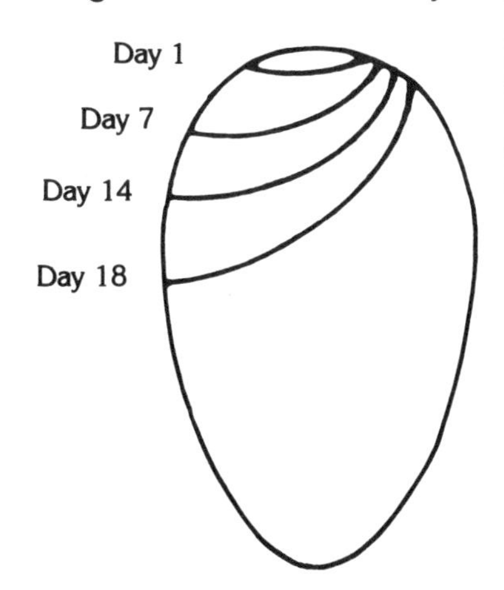

Increasing Air Space in the Hatching

(Taken from Scholtyssek 1962, *Die Geflügelzucht-Lehre*)

Faults in Hatching with Their Corresponding Incubation Factors

The following table outlines hatching faults and suggests the probable causes.

Hatching Fault:	*Reason:*
Infertile eggs after the first candling	Too few roosters; infertility in one sex; sexes not yet accustomed to one another; eggs too old
Clear eggs after the first candling, blood ring or small embryos when opened	Incubation temperature too high; too much cooling; breeding stock diseased or poorly nourished
Dead embryos	Incorrect incubation temperature; too little oxygen in the incubator; poorly fed parents
Chicks develop but not viable, remain in egg	Faulty turning; faulty temperature; inherited disposition
Egg pipped, chick dead in shell	Average humidity or temperature too low; temperature elevated excessively for a short time
Chicks stuck in shell	Eggs dried out; too low humidity during hatching
Stuck chicks	Temperature too low; humidity too high
Yolk sac not absorbed	Temperature too high or too fluctuating; humidity too low
Small chicks	Eggs too small; incubation too dry or too hot
Large, weak chicks	Incubation too moist or too cool; poor ventilation
Dead chicks with foul odor	Navel infection
Chicks with sparse down-feather formation	Temperature too high, humidity too low
Premature hatching, chicks with bloody navel	Temperature too high
Delayed hatching	Temperature too low
Crippled chicks	Crossbeak is hereditary; deformed toes are a temperature fault

Taken from S. Scholtyssek, *Handbuch der Geflügelproduktion (1968).*

A flotation test indicates which eggs are developing.

embryonic development usually ends with the death of the too-rapidly growing chicks. This happens with an excess temperature of 1.5° C. If it lies below this limit, weak chicks often hatch; very many get stuck in the egg during the hatching phase or die beforehand. A brief cooling of the eggs at room temperature (about five minutes) has a beneficial effect on hatching with waterfowl, since by this means the metabolism of the embryos is stimulated. A cooling of the egg of this kind corresponds to the space of time in natural incubation during which the broody hen leaves the nest to feed.

Humidity

During the incubation process the egg experiences water loss. This is based on the process of evaporation. So that the development of the embryo is not disturbed, the degree of evaporation must be held within limits favorable to metabolism. These limits are ensured by the relative humidity; in addition, it promotes gas exchange between the

Average Values for Artificial Incubation

Fowl	*Temperature*		*Relative Humidity*		*Turning*		*Candling*
	Day	*°C*	*Day*	*%*	*Day*	*Frequency*	*Days*
Chickens	1–17 18–21	37.8–38.0 37.0	1–19 20–21	55–60 80	1–17	4 times	6, 17
Turkeys	1–22 23–28	37.5–37.8 37.0	1–24 25–28	55–60 80	1–24	4 times	9, 22
Geese	1–16 17–27 28–30 (from the tenth day, cool twice to room temperature)	37.5–37.8 37.3–37.4 36.5–37.0	1–28 30	60 80	2–25	2 times 120°	10, 25
Ducks	1–22 23–28 (from the tenth day, cool twice daily)	37.8–38.0 37.0–37.5	1–22 23–28	55–60 80	2–22	2 times 180°	7, 14 22

Taken from S. Scholtyssek, *Handbuch der Geflügelproduktion (1968).*

egg and the environment. In the first part of incubation (first to nineteenth days with chickens) the humidity should be about 60% and should be about 80% in the hatching phase (twentieth to twenty-first days of incubation). Depending on diet — whether very protein-rich or not — it is also beneficial to allow completely dry incubation during the first week so that a high water content evaporates. Low humidity should not prevail for too long, however, since otherwise the egg will lose too much liquid. Usually the chicks will then no longer free themselves from the eggshell. If, on the other hand, the humidity was too high, then the air space remains small, and the egg white and yolk are not completely reabsorbed, so egg white remains in the egg. The chicks then easily remain stuck and can drown in the egg white. Often they are not able to draw in the yolk sack, which provides them with a nutrient reserve in the first days of life. Chicks that do hatch have a weak constitution. On the other hand, a diet to rich in sprouted grain will lead to an especially heavy egg white concentration in the egg. As a consequence, the chicks again drown in the egg or remain stuck. As a remedy, the previously mentioned dry brooding during the first third of incubation (and at times somewhat longer) is recommended. The size of the air space can be used as a criterion for determining the proper humidity. On the eighteenth day of incubation it should occupy about one third of the volume of the egg. The accompanying diagram shows how the air space should enlarge.

The air space can be seen by candling the egg. For this purpose special candling lamps are available on the market. We can, however, also easily build one ourselves. We build a box out of wood in which there is room for a light bulb and socket. In the lid of the box we saw an egg-shaped hole. If we deal with eggs of different sizes, then cardboard templates provide variable hole sizes. We place the egg on the appropriate hole and turn on the bulb, and we can clearly see the air space.

In the same manner we check to see whether or not the eggs are fertile. For this purpose we take the eggs out of the incubator on the appropriate candling date (see table) and candle them. With the chicken egg, for example, after seven days we can recognize the beating heart as an upward and downward moving point about three millimeters in size. In addition, blood vessels run through the interior of the egg in its vicinity. (If, on the other hand, the egg is clear as glass after seven days, it is infertile.) On the seven-

In this type of incubator the eggs lie on rollers.

Eggs in a well-formed nest, ready for 21 days of incubation by a brood hen.

teenth day of incubation we candle the egg once more. Now the contents of the egg must be almost completely dark except for the air chamber. If the egg is relatively clear, or if only a part is dark, then the embryo has died. Dead and infertile eggs are removed from the incubator.

If we wish at a later date to check whether the eggs have died, chicken eggs should be soaked in a water bath with a temperature of 38° C. on the nineteenth day of incubation. They will float on the surface if the air space is enlarged. If the eggs sink, then the breeder used too high a humidity and the air space stayed too small. The eggs floating on the surface clearly show twitching, hopping, and rotating movements caused by the chicks moving inside. Eggs that do not move have died.

Turning

Turning the eggs daily is very important for development, since otherwise the embryo could stick to the shell membrane.

If disturbances in embryonic development occur, then the cause could possibly lie in too little turning. Turning the eggs should take place at least twice a day with chicken, turkey, and guinea-fowl eggs. Eight turns a day would be ideal. Turning is omitted at the time of hatching, since it impedes hatching.

Air Supply

Supplying oxygen to the eggs is very important. If a constant supply of oxygen is lacking, then the carbon-dioxide concentration increases through embryonic metabolism and leads to the death of the unborn

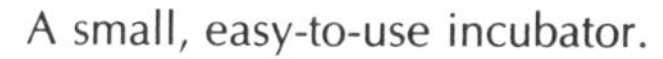

A small, easy-to-use incubator.

The well-appointed breeder will have an incubator of this sort.

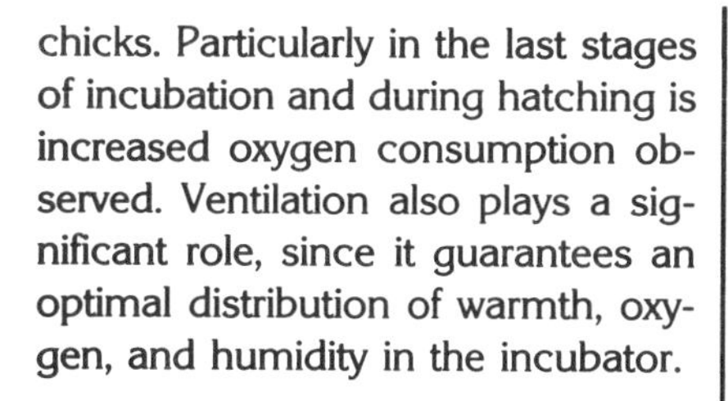

chicks. Particularly in the last stages of incubation and during hatching is increased oxygen consumption observed. Ventilation also plays a significant role, since it guarantees an optimal distribution of warmth, oxygen, and humidity in the incubator.

Incubation Failures Related to Hatching Factors

The accompanying table gives indications of the outward appearance and causes of incubation failures.

To be sure, the novice breeder will at first use a broody hen for hatching, or have the eggs hatched at the installation of another breeder. But as time goes by, the desire to own an incubator will arise. Manufacturers regularly advertise their equipment in specialist publication. The price range varies widely, so there is a model for every taste and pocketbook.

Marking the Chicks

If the chicks have hatched, then we can safely leave them in the incubator one more day so that their down dries and they develop leg coordination.

Before we transfer the chicks to the rearing coop we mark them so that later we will know exactly which chicks are descended from which hen. For this purpose we must of course have carried out the trap-nest inspection already discussed, and during the hatching phase the eggs of each hen must have been kept separately. The more expensive incubators have so-called hatching lattices, which are divided into separate areas that can be used for this purpose. But one can also build dividers of this kind for any incubator using plywood slats.

We mark the young of chickens, guinea fowl, and turkeys by means of toe-skin punching or chicken wing bands; we use the latter exclusively for ducklings and goslings. We can purchase toe-punch pliers and chicken wing bands from poultry supply houses.

With toe punching we make a hole in the foot web all the way back between the toes. A hole farther forward easily tears, and ultimately does not look as good. The perforation itself causes only minimal pain; often the small wound does not bleed at all. Since we can punch in various positions (for example, right foot right or left, or left foot right or left), numerous combinations are available.

In using chicken wing bands we pierce the skin of the wing with the point of the wing band in the region of the secondaries. With fragile-winged birds we do not undertake this until several days after hatching, since the wing band will otherwise tear out easily. Until banding can be done, we mark the chicks with a colored spot on the down. A waterproof felt-tip pen, with which we place a mark on various parts of the body to indicate which hen the chicks are descended from, is suitable for this purpose. The use of different colors increases the possibilities.

Rearing by the Broody Hen

If the broody hen — it does not matter which poultry group she belongs to — has hatched her chicks, then we transfer them to a rearing pen. For litter we place dry sand, short straw, or fine leaves in the coop. Feeding takes place according to the same principles as with artificial rearing (in which the feeding methods are discussed). When using a broody hen, one has hardly any problems with the chicks, since she is devoted to the well-being of the little ones. With geese and particularly with ducks, opportunity for bathing is necessary. One should let the little ducklings outside only in good weather with the duck or the adopted mother, since at first the oiling mechanism does not yet function and they can quickly catch a deadly cold. Turkey chicks need considerable warmth, especially in

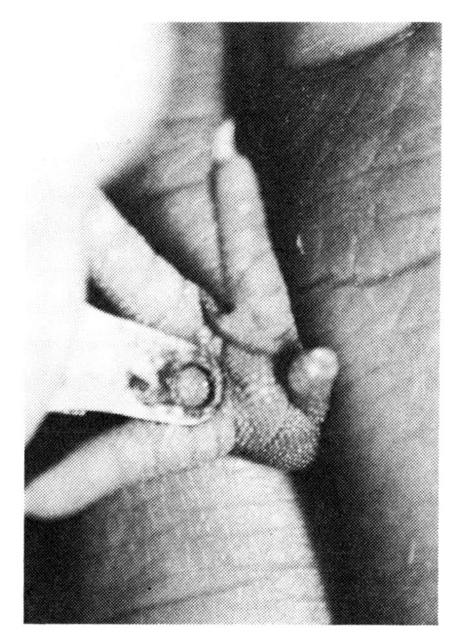
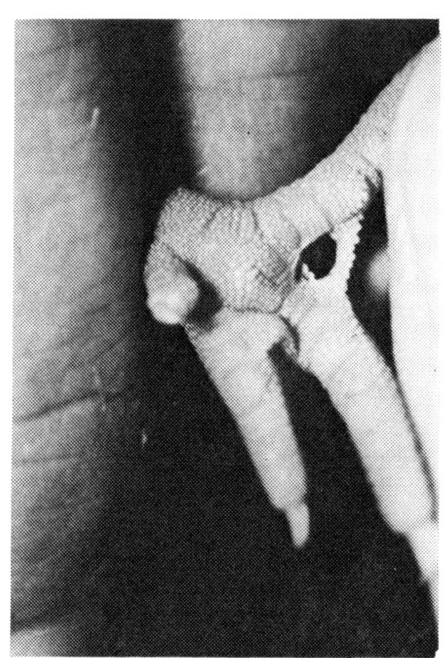

Toe punching provides a certain way to indicate a bird's ancestry.

the initial phase; therefore, one should at first keep them with their mother in the pen for a fairly long time. With goslings, a separate pen for the mother with her goslings is often unnecessary, since, because of their pronounced "social life-style," geese can rear the little ones in the flock without much difficulty.

Artificial Rearing

Rearing Chickens

Rearing Area, Space Requirement, and Temperature

The artificial rearing of chicken youngsters begins before the first chicks have hatched. We disinfect the brooder (a cabinet-like box with several rearing compartments and a floor of wire mesh) for hygienic reasons; we check that the heating apparatus is operational. Finally, the brooding temperature is set on the desired value. The rearing area should have been heated to the necessary temperature at least one day, or better yet two days, before the chicks are put in. The initial temperature is approximately between 32 and 35° C. during the first week. To regulate the temperature exactly the breeder must observe his chicks. If they huddle close together under the heating lamp, the temperature can safely be raised slightly. If they, however, stay at the edge of the brooder — far away from the heat source — this is a sign that the chicks are too warm, and one lowers the thermostat slightly. If one has established a comfortable temperature in this manner, then every week one reduces the temperature by two degrees C. After about eight weeks we can turn off the heat, provided the birds have been gradually acclimatized to life without the heat source from the sixth week on. This is done by turning off the heat periodically: at first during the warmer midday hours, then during the entire day, and finally also at night.

The selection of heat sources is diverse. Infrared lamps, dark infrared radiators, heating coils and rods, and many more are available. The type of device the breeder ultimately decides on is often determined by the breed. The growth of the comb is stimulated by the use of an infrared radiator; however, it often leads to feather picking. In relatively dark rearing areas heated with coils or rods, feather picking virtually never occurs. However, chicks raised under the "darker" conditions do not grow as rapidly as with an extended

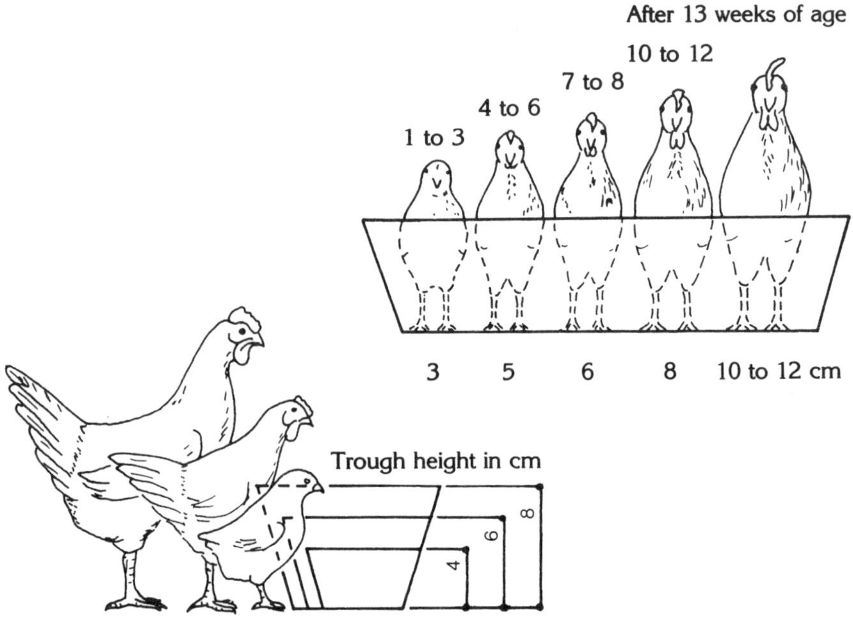

Changes in Trough Height and Length with Increasing Age
(Taken from Pingel 1981, *Kleintiere richtig füttern*)

light period. But it is known that unforced growth is not exactly a disadvantage for the total development of the organism (unless one wishes to raise the chicks for meat).

Whoever does not own a brooder — which is recommended to everyone, since it guarantees very hygienic conditions because of its wire-mesh grating and requires only minimal work for rearing — must prepare a chick pen. This consists of a smaller sleeping and resting area and a larger activity and feeding area. Chopped straw, wood shavings, and sand have proved effective as litter. If possible, one should also install a wire floor in this case, through which the droppings can fall. The feed and water containers are placed in such a way that they can neither be soiled by litter nor droppings. Wet areas — particularly near the water container — must not be allowed to develop, since microorganisms can multiply there very easily. Infectious diseases are often the inevitable result.

Whoever possesses neither a rearing cabinet nor a chick pen should furnish a suitable area in an empty coop, in which one part is used as a roosting area — with the heat source — and the other part serves as a place to feed. The total area should be limited, at least in the beginning, to a round to oval area by means of corrugated paper, which prevents cowering in a corner of the coop. Once the slightly larger chicks have grown accustomed to the heat source and the coop, we can remove the barrier. Of course, there must not be any draft in the coop.

It is essential that overcrowding does not occur. Approximately 15 chicks of a medium-weight breed require one square meter of space. A slight overcrowding is allowable at first, but should be avoided in the long run. Weak chicks should be culled. Because of their meager growth they lag behind and will seldom become usable production, breeding, or show birds. The feed consumed does not bear any relation to their later value. Such chicks also are often carriers of latent (hidden) diseases and so present the danger of infection for the remaining chicks. If one culls them, then one creates better living and keeping conditions for the vigorous chicks. Chicks that are otherwise healthy but have deformed toes or other malformations should also be culled.

Rearing Feed

We first feed the chicks in the brooder or in the feeding area of the chick pen. This is best done initially not with a feed trough but with a flat tray. Here the chicks can learn to peck. Once they recognize the feed and have learned the pecking movement, they will quickly learn to feed from the trough. One should allow about 8 grams of feed in the first week and 14 grams in the second week per bird per day; by the eighth week the feed requirement will have risen to 53 grams. The feed containers must be large or long enough that no jostling and scuffling occurs, since the weaker chicks are always pushed aside. It is most practical to feed the chick feed prepared by feed companies, since it is optimally suited to the chicks' needs. The chick meal, which we feed until the sixth week, can be purchased in mash and pellet form. The pellets satisfy hunger quickly, and the feed loss through the pecking and spilling of feed from the trough is minimal. The quick satisfaction of hunger, however, often leads to boredom, which can end in unpleasant feather picking. Feeding mash presents a diversion because of the longer feeding time, though the feed lost is quite considerable. Therefore, the feed troughs should only be filled a third full. At the same time, the trough height and length must be suited to the age of the birds.

Drinking Water

Lukewarm drinking water is best for chicks. At first it is often necessary to show the chicks the water by dipping their beaks lightly in it. Tea supplements in the first two weeks are helpful. Of course we only use

In a brooder, the chicks—only a few days old—should have their feed in a shallow container.

medicinal herbs, above all camomile. Fresh tea must be given daily. A vitamin supplement promotes the health, growth, and development of the chicks. Abundant feedings of greens (stinging nettle, dandelion, chickweed, and so forth) also supply the chicks with vitamins and minerals. For optimal digestion the chick must be given quartz sand or grit. Whoever does not wish to use prepared feed can feed so-called chick groats (slightly coarse-ground grain). Since this only contains about 9% protein, however, the food value must be increased through the addition of meat meal, fish meal, alfalfa meal, dry yeast, and so on. Chicks need an 18% raw-protein content for growth. However, some chick starter feed available on the market may speed the growth of the small chicks too much.

The regular inspection of the chick flock is essential. Often balls of droppings form on the feathers of the hindparts or deposits of dirt collect in the toe claws when chicks are reared on litter. Softening in warm water and careful removal of the dirt is the remedy.

Rearing Ducklings

Rearing Area, Space Requirement, and Temperature

A dry and draft-free rearing area is necessary for raising healthy ducklings. In addition, the area should be perfectly hygienic and preferably should be warm from the ground up. If the floor is cold or even wet, then the ducklings contract very serious colds even though the infrared radiator is turned on.

To prevent a wet floor, it is best to divide the duckling rearing area into feeding and sleeping compartments. The feeding compartment should make up about a third of the total area. To guard against wetness, the feeding area consists of wire mesh with a basin placed underneath to catch water. The feed and water containers stand on the wire mesh itself. Here splashed water and wasted feed cannot wet the litter. The droppings, which are primarily discharged during feeding, do not soil the sleeping area in this set-up. The rest of the rearing area is the sleeping compartment. For reasons of warmth it should be constructed of wood. It is practical to install a wooden slat between the sleeping area and the rest of the rearing area, so that the litter from the sleeping area is not carried into the feeding compartment.

Here again, the brooder with its hygienic floor of wire mesh is ideal. If one places the little ducklings in a middle rearing drawer so that heat also rises from the rearing compartment below, then the ducklings have "floor heating" and are also warmed by the brooder lamp.

The air supply in the rearing area is also important. Since ducks expel the majority of the water they consume through respiration as water vapor, the coop has a tendency to become damp. A clever and carefully thought-out ventilation system is the answer. Fresh air must be able to enter the coop, become warmed in the coop, and then leave it again as moisture-saturated air. A draft must of course not be allowed to develop, since otherwise the ducklings will soon suffer from colds and may die under certain circumstances. If the humidity in the rearing area is low (a hygrometer in the coop will supply this information), then a temperature of only 24° C. is sufficient in the initial phase. Higher temperatures are not suitable because of the need to acclimatize the ducks to lower temperatures. In the following weeks one reduces the temperature by three degrees C. a week, and after two weeks one should allow the ducklings to go outside occasionally in good weather.

If the ducklings signal by constant peeping that they are cold, then we raise the temperature under the heat source (at head height) to 32° C. From the sixth to the tenth day of life, however, we should reduce the temperature to 24° C.

When choosing the litter, one must turn to relatively coarse material (assuming that rearing does not take place in a brooder), since ducks, with their instinctive greed, try to eat everything. If the indigestible litter is small and easy to take up, then they will swallow it. If it sticks in the crop, it usually leads to the death of the ducklings. Long, coarse wood shavings are well suited as litter. If one allows the ducklings out of doors in good weather at a relatively young age, they should encounter clumps of grass or clean sand.

Rearing Feed and Water

As feed one should initially offer barley and corn groats as well as cooked rice (in equal amounts). If one mixes the feed with finely shredded greens (for example, stinging nettle) and grated shrimp, then one has an excellent rearing feed. The mixture is offered to the ducklings in a moist-crumbly state. If the feed is dry or too liquid, then they quickly lose their appetite. If the feed is too moist, one runs the danger that the down will become smeared and protection against chilling will no longer be provided. As an alternative one can offer rolled oats and breakfast rolls soaked in milk. A supplement of greenfood is then obligatory. The most suitable are stinging nettle, dandelion, spinach, and of course the valuable duckweed, which unfortunately can no longer be found everywhere. Whoever wants to take the easy way out feeds complete feed for ducklings with 18% raw protein until the sixth week. As a substitute, chicken chick feed (not medicated!) with the same protein value is also suitable.

Ducklings are fed every three hours. Long troughs, at which all the ducklings comfortably have space, are the most suitable. If the troughs are too short, then jostling will occur and the stronger ducklings will of course get their way. The consequence is unequal growth. The water containers are placed as far as possible from the feed, because otherwise too much feed, which has stuck to the bill, will be carried into the water. As far as possible, the drinking containers should be enclosed on top, since the ducklings will otherwise splash, thereby fouling the water and wetting the surroundings inordinately. One gives them a separate container for bathing. After bathing, the ducklings are often soaked. This is not bad if the birds can quickly reach the heat source. There they dry very rapidly while preening. Since their preen gland is not fully functional at first, they can become thoroughly soaked. In this phase they absolutely require a source of heat, since otherwise they are very susceptible to colds. The bathing containers must be readily accessible and must also be easy to climb out of, since otherwise there is the danger of drowning.

Rearing Goslings

Rearing Area, Space Requirement, and Temperature

Goslings hatched out in the incubator should in principle be kept like ducklings. In the first three days the goslings require a temperature between 29 to 31° C. From then until the seventh day they need 28 to 30° C., and then 25 to 27° C. up to the twelfth day. From the thirteenth to the eighteenth day, 22 to 24° C. is required, and from the nineteenth to the twenty-first day, they need 18 to 21° C. Subsequently the temperature is kept at a constant 18° C.

Rearing Feed and Water

For feed one offers finely shredded stinging nettle and dandelion leaves which are mixed with barley groats and wheat bran. Gosling starter feed is beneficial in the initial phase (about two weeks). After that, provide duckling feed or chick crumbs. The feed should be offered in a moist-crumbly state. Supplemental feedings of rolled oats, egg, and breakfast rolls soaked in milk are readily taken. With increasing age, in the evening, in place of the initial grain groats, provide simple grain since it keeps well overnight. Drinking troughs with fresh water are very important. They must be constructed in such a way that the birds cannot enter them. The larger the birds become, the more they should be able to seek out their own food. Large meadows are the most suitable for this purpose but are not available to most breeders. The feeding of large amounts of greenfood serves to some extent as a substitute. Taking in sand promotes digestion during the rearing phase.

Rearing Turkey Chicks

Rearing Area, Space Requirement, and Temperature

Turkeys are very delicate during rearing. At least in the first few weeks, keeping them on wire mesh is ideal, since by this means the dreaded blackhead disease especially is prevented. If one keeps the birds on the floor of the coop, then a layer of chopped straw or wood shavings, which should have a depth of about 15 to 20 centimeters, is recommended. The room temperature should be about 30° C. in the first week, and then should be reduced by two degrees C. each week. In the seventh and eighth week we maintain the temperature at a constant 18° C., and subsequently — until the twentieth week — we lower the temperature to 16° C. Under the heat source the temperature is 35 to 38° C. at the height of the chicks' heads in the first eight days and is lowered by three degrees C. each week. In the seventh week the room temperature and the temperature of the heat source should lie between 16 and 18° C. The relative humidity should be between 60 and 75%. Drafts should be avoided under all circumstances.

Rearing Feed and Water

During the first six weeks we use complete feed for turkey chicks with a raw protein content of 26%. Occasionally one can also feed turkey starter feed with 28% raw protein. The prepared feed contains special supplements that help protect the susceptible chicks against illnesses. Therefore, one should not mix one's own feed but should feed the special turkey feed instead. Greenfood is also highly recommended in this case, as is a vitamin supplement in the drinking water and a supplementary feeding of minerals.

After three weeks the birds may go outside for the first time if the weather is warm and dry. Otherwise they are kept like chickens; the rearing area must not, however, be too large, since turkey chicks often cannot find their way back to the heat source.

Rearing Guinea Fowl Chicks

Guinea fowl chicks are kept in the same way as chickens, and they also receive the same feed.

From the "Half-grown" to the Mature Bird

Keeping Young Chickens

When the youngsters have grown out of the chick stage after about six weeks and enter the stage of the pullet or cockerel, then special keeping, feeding, and care are necessary. For only through healthy development and satisfactory growth does the breeder lay the cornerstone for his show, production, and breeding birds with the growing generation.

At the age of four to six weeks one can determine the sex of chickens. A pronounced comb is a sure sign of a maturing rooster; the rooster also usually differs from the hen in body form and plumage coloration. At this point in time the young roosters should be separated from the pullets. Separate rearing has advantages for both sexes. Young hens are not pestered by the maturing roosters and thus develop better: their development takes an uninterrupted course. For young roosters it is beneficial if they do not begin mating too soon, since their bodily development could be negatively affected in this way.

Keeping Pullets

But not only for this reason does rearing the sexes separately offer advantages: the possible feeding methods also have definite merits. The hens are switched from chick mash to grower mash. This change must not take place too rapidly, since otherwise disturbances in growth can occur. Therefore, one should proceed cautiously. During the change-over phase one gives the young hens three parts chick mash to one part grower mash during the first week. In the second week we feed chick and grower mash in the ratio of 1:1, and then in the third week one part chick mash and three parts grower mash. From the fourth week on, pure grower mash is used. This contains about 15% raw protein. From the thirteenth week on, we can also feed a grower mash with about 13% raw protein, since this promotes a slow and harmonious growth of the birds. Following the same methods as with the chick-grower mash change-over, we switch from grower mash to layer feed about three to four weeks before laying begins. With young roosters it is sometimes beneficial to leave out the grower-mash step and to switch immediately to layer feed. The higher protein content of the feed helps to force growth in the slower-developing roosters. Nevertheless, one must not forget that no coccidiostats are contained in laying hen feed, so that the young roosters do not obtain the same prophylactic (preventive) protection as young hens. A separate, planned prophylactic coccidiosis treatment via the drinking water is appropriate in this case. If one also omits the grower-mash feeding step with young hens, then one runs the risk that the hens will begin laying too early, which of course is not good for the birds.

Neither young hens nor young roosters should be given too little calcium, since otherwise deformed toes can result.

We also distinguish between mash and pellet forms of grower meal. The pros and cons have already been treated.

Ten youngsters require a trough length of at least 40 centimeters, although 80 to 100 centimeters is ideal. If the absolutely minimal space is not offered, then — as with chicks — variable rates of growth result, since the stronger birds always displace the weaker ones. The feed troughs must not be too wide or too narrow. In the second case comb injuries can easily occur, and in the first case the chickens often stand in the trough and scratch. This often leads to high feed losses, which increases the cost of rearing. Besides feeding in troughs, feeding with automatic feeders has also proved effective. With these the amount of feed dispensed can be regulated continuously, since there are various control mechanisms available. It is certainly obvious that the feed containers must be set up so that they do not permit the entry of litter.

The drinking-water supply is also very important. The water must be fresh and cool. Chickens drink stagnant water only reluctantly. Consequently, their development lags behind the norm, since the body, which is generally known to consist of over half water, receives too little of this essential elixir. Before filling the container with fresh water, it should be cleaned with a small sponge or a brush.

In addition to proper feeding, accommodations are also important. In general, the coop must not be allowed to become overcrowded. With increasing age the youngsters require the same amount of space as adult chickens. If one has a high, but acceptable population density, then the litter must be changed frequently. Checking the ventilation is essential, for this ensures the supply of oxygen and keeps the litter dry. Drafts lead to colds, which naturally have a negative effect on the development of the poultry. Concerning perch space, one allows 15 centimeters for light breeds and 22 to 25 centimeters for medium-weight and heavy breeds. The coop's roosting compartment is best placed away from the window with young poultry as well, while the feed and water containers are located in an area flooded with light.

Delimiting various spaces is as important with the run as it is with the coop. Sand piles, shrubs, perennial flowers, or artificially erected wooden fences divide the run into several areas. Each of these zones has a high intrinsic value for the chicken. Here it can move from one area to another, and thus get the impression of having an inestimably large territory. This is very important psychologically and so benefits development. In addition, a lower-ranked chicken engaged in a fight can quickly disappear from the field of view of the stronger one and is not constantly exposed to attack. This is particularly important in housing young roosters. Because of the structuring of the enclosure, the run provides many shaded areas which in summer protect against the strong sun. A roofed dust bath is indispensable for the birds' health and hygiene.

For youngsters, the run should be unoccupied by chickens for a fairly long time. If the surface of the run is covered with droppings, these should be removed, since they represent a virtually ideal "home" for disease germs. If the soil of the run is turned over, quicklime should be worked in, since this destroys the pathogens and speeds up the decomposition of the droppings.

Keeping Young Roosters

With young roosters we should make sure that they do not fight too much. In fighting, a comb point is often lost or the heads become so encrusted with blood that a purebred-poultry breeder can no longer show the birds at an exhibition. In addition to the structuring of the enclosure, which is supplemented with flight perches with hanging sacks which end about 15 to 20 centimeters above the ground, one also puts so-called "chicken blinders" on the most aggressive young roosters. One places these on the rooster's upper mandible and fastens them with a pin which is pushed through the nostrils. In so doing the nasal septum must be pierced, which is difficult to do with the flexible pin supplied with the blinders. Therefore, it is best to first pierce the nasal septum with a large pin or similar pointed object. Then the pin supplied with the blinders can easily be pushed through. It is best to carry out the entire procedure with another person, holding the bird while the other puts on the chicken blinders. Nevertheless, its effect is often not what is promised, and an attack takes place despite the limited forward field of vision.

Another possibility for stopping fighting lies in putting on so-called leg chains. For this purpose, some type of band is placed on each leg of the most aggressive roosters; the two bands are then tied together with a string. The string is long enough that the rooster can still walk satisfactorily, but long strides are prevented. Thus it can no longer run and chase weaker roosters. The string must never be tied directly to the legs, since they will be drawn closer and closer together by the rooster's constant movement, which will have fatal consequences for the bird.

If a purebred-poultry breeder has taken his best rooster to a show and then introduced it back into the rooster flock, a fight is virtually unavoidable. The birds, which have since become strangers, see one another as rivals, and a fight for position in the hierarchy will take place. If one is lucky, it will be only of short duration and without serious injuries. Often, however, one is not so lucky. In order to successfully prevent these fights, there is actually only one effective measure: caging. In other words, we do not return the bird back to the rooster flock at all, but instead we keep it separately in a cage. Since one usually exhibits several roosters, one of course will need a row of cages.

This cage must be large enough that the bird has sufficient room to move. A size of at least one square meter is appropriate. If the space for a cage is not available, then it is better to put the rooster back in with the rooster flock, even at the risk of injuries, since our foremost goal should always be keeping poultry in a humane manner. (Incidentally, the empty rooster cages are also well-suited in spring as brooding places for broody hens, since there they can regulate their own daily activity.)

While roosters still stay together in the rooster coop as adults, pullets hens are placed in the laying coop about one month before the start of laying. After the transfer, the young hens remain in the coop for about three days. During this time they become accustomed to their new surroundings and the remaining old hens from the breeding flock. Only after that should the birds be allowed in the run. In the evening the breeder should make sure that the birds roost on the raised perches. So that the young hens do not undergo a neck molt (the neck molt is a partial molt that only affects the neck area), they must have constant and extended periods of light. The duration should be about 14 hours. If fall and winter are approaching, then the breeder must compensate for the shortened day by means of artificial lighting. If this measure is carried out in time, the disagreeable neck molt, which is often a great nuisance to the purebred-poultry breeder who wants to present his birds in exhibitions, will not occur.

Banding

The banding of the birds also takes place during the rearing phase. An exact time for this cannot be given, but in general it lies between the eighth and twelfth week. Various types of bands can be obtained from poultry supply houses. Exhibition rules may stipulate that the birds be banded specially in order to enter.

If one has placed chick markings on the birds, then these can be removed after banding. One must, of course, keep notes on which youngster carries which ring; otherwise, the marking and the previous trapnest inspection would have been in vain.

In banding, one pushes the three front toes through the ring and then presses — with chickens — the hind toe slightly toward the leg and pushes the band over it onto the leg. With five-toed breeds we do not bend back both the hind toe and the fifth toe.

If one has allowed the most favorable date for banding slip by, then it is often not yet too late to band. However, the use of gentle force will now be necessary. First of all we massage the foot pads and the toe

Various types of bands are used to mark purebred poultry.

joints by gently pressing and stretching them. Then we apply petroleum jelly or soap to the toe joints and try with continuous turning to slide the band over the ball of the foot. One should make sure that the bird's leg scales are not injured in the process. Sometimes one can also bend the band slightly to make it conform to the anatomy of the foot. Then it will be impossible to turn the ring, however. Instead we must push the band upward. While pushing we constantly press the individual scales under the band with a fingernail, since they otherwise break off. Nevertheless, it should never get this far. A model breeder has always ordered his bands early, and puts them on when they can be slipped over the foot without problems.

It is very important to give the correct size when ordering the bands. With the exception of guinea fowl, and some ducks and geese, chickens need bands of different sizes depending on the sex. The required band sizes for each breed are given in the poultry standards and in catalogues.

Keeping Young Ducks

For housing young ducks one allows one square meter of coop area for about five ducks. Since ducks are very lively, one leaves them in the run constantly; only at night and in very bad weather should they be offered a coop. For better feather growth, running water is necessary. The formation of puddles in the run should be prevented. A grassy run or ground covered with clean sand are ideal. The coop litter, which consists of short straw, wood shavings, or sand, must not be allowed to become wet. For this reason fresh air must be provided constantly. Otherwise the coop is furnished as for mature ducks.

With ducks the sexually determined breaking of the voice appears at five weeks. Drakes have a quiet, hoarser call, while ducks quack and gabble loudly.

Starting in the seventh week we switch from duckling mash to complete feed for young ducks, which contains about 13% raw protein. If a low percentage mash of this kind is not available, then we can also feed one with a higher percentage, but must reduce it according to the values for the nutrient ratio (as explained in the chapter on diet) using other foods. The switch from duckling to young-duck feed takes place as with chickens. With the start of laying or about four weeks before, we change over in the same manner to complete feed for breeding ducks (16% raw protein) or carry out an already selected alternative feeding.

Keeping Young Geese

Since geese are grazing animals, youngsters too need a large run. For the coop the same requirements as for the duck coop should be met. We allow three young geese for one square meter of area. If there is no danger from predators we can also leave the young geese outdoors at all times. However, they are then provided with a roofed shelter to protect them against bad weather.

As a foundation one offers them the same feed as ducks. As with all waterfowl, an abundant variety of greenfood (especially finely shredded stinging nettle) enriches the menu. With heavy goose breeds a protein-rich diet is also beneficial.

Opportunity to bathe — as with ducks — has a favorable effect on development.

Keeping Young Turkeys

Although turkeys should have a temperature of 16° C. until the twentieth week, one can leave them outside all day after about ten weeks, since by this time we usually already have spring-like to summery temperatures. Turkey youngsters need lots of run space; then they roam about widely and search out their own food (insects, snails, worms, plants, and so forth). The coop should provide good protection against the effects of weather. Perches make the youngsters' instinctive roosting possible.

From the seventh week on — as with feeding chickens — one changes over from protein-rich turkey starter feed to young-turkey feed with 15% raw protein. At the start of laying we feed the complete feed for breeding turkey hens with 16% raw protein, or an alternative, self-prepared breeding feed. Greenfood complements the diet here as well.

Keeping Young Guinea Fowl

Guinea fowl are housed and fed like chickens. They are also very grateful when given abundant amounts of greenfood.

DISEASES AND PREVENTIVE MEASURES

Prevention of Disease

Prophylaxis is of fundamental importance with poultry. It is attained through proper accommodations, care, and diet.

A reduction of contact between the birds and their droppings is essential. For this purpose — as was previously discussed — we install a droppings pit, which is covered with wire mesh, under the perches. If possible, we always rear chicks on wire-mesh floors. The feed and water containers are set up such that they cannot be soiled by litter or droppings. The litter itself must always be clean and dry, since otherwise pockets of infection can easily develop.

We regularly disinfect the coop. We should always soak a few pieces of onion or garlic in the drinking water, since their active agents protect against some illnesses. But these natural remedies can also be administered just as effectively through the feed or by individual force-feeding. To do this we take the bird in our hand so that its belly rests on the palm. Our index finger is located between the legs, while the thumb and the remaining fingers hold the legs tight. Even if the wings flap, it cannot escape from us, since we of course have it under our control. After we have the poultry in our power with this universal grip, we open the beak and put the pieces of garlic or onion in the bird's gullet. We may not hold turkeys in this manner, since they could quickly go into a "shock molt." (With a shock molt caused by fright, the birds abruptly lose their feathers on the areas of the body that were touched.)

Diseases

Embryonic Diseases

INFECTIOUS EMBRYONIC DISEASE

The white of the egg of course possesses bacteriostatic substances that prevent infection, but this mechanism only functions if the breeding birds are cared for properly. Incorrect diet, poor coop hygiene, and deficiencies in the vitamin supply reduce the self-defense mechanisms of the embryo in the egg. For this reason, infectious embryonic diseases will appear especially in breeding flocks that are not given suitable care. Furthermore, the egg's protection against bacterial attack decreases with the length of time the egg is stored. For this reason one should — if possible — always use fresh eggs for hatching, because with older eggs one runs the risk that they will die during incubation.

The infection can occur in two ways: either during egg formation in the hen, or through the eggshell during storage and during the incubation process. One speaks of an "endogenous" or an "exogenous" infection.

With the endogenous infection the disease is transmitted from the mother to the egg. As a rule, it is a matter of typhoid, tuberculosis, Newcastle, or pullorum disease, and similar bacterial and viral diseases. This means nothing more than that a hen with a latent illness transmits her disease to the offspring, which then are not — or are only conditionally — viable. For this reason only healthy birds belong in the breeding flock. Poultry which were sickly at one time but became healthy again should not be used for breeding, since the disease can remain — if only latently — in the bird. The danger of infecting the next generation would be too great.

With the exogenous infection, on the other hand, the viruses or bacteria enter the egg through the shell during incubation or during the storage of the hatching egg. The pathogens penetrate the protective barrier of the cuticle of the surface of the eggshell and overcome the embryo's defense mechanism. Exam-

In cases of serious illness, the breeder must resort to medicines and possibly seek veterinary advice.

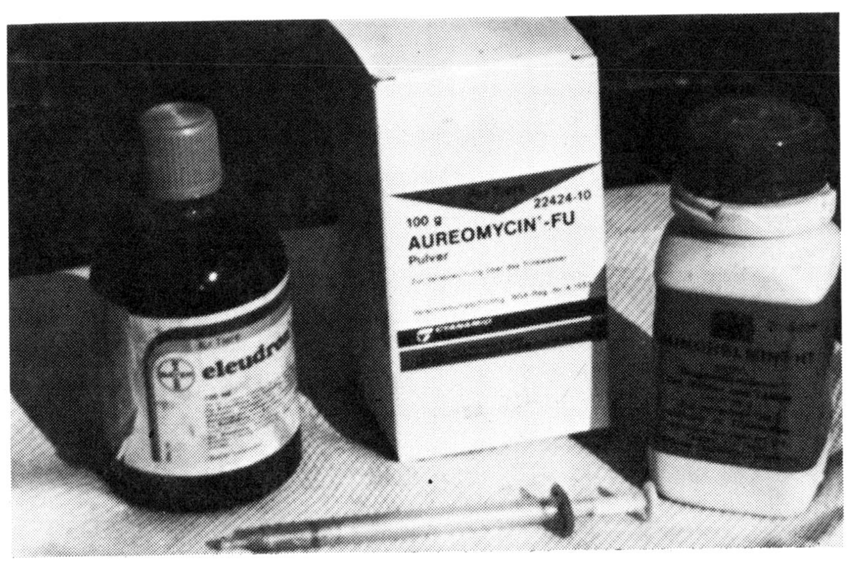

ples of exogenous pathogens are fungi, paratyphus, and similar bacteria. If embryos are afflicted with exogenous infections, then the breeder either has an unhygienic coop, faulty egg storage, or an undisinfected incubator. Particularly with this device, in which the moist, warm air promotes the growth of large numbers of bacteria, the greatest possible hygiene is necessary. If the bacterium has penetrated the egg shell, then it increases its numbers inside the egg. The egg is then the nutrient medium, and the incubation area supplies the favorable external conditions. The embryo dies in the egg. A specific time of death cannot be given, but it usually does fall in the second half of incubation. With some endogenous infections, such as tuberculosis, the chicks do hatch, but the latent disease manifests itself at a later age, which presents the breeder with an apparently unsolvable puzzle. That the chicks were already condemned to death during the production of the egg or during incubation is usually overlooked during the search for the cause.

To protect against such unwelcome surprises it is important that one keeps his birds free of infectious diseases. Sickly birds do not belong in the bird flock — and definitely not in breeding. But attention should also be given to the greatest possible cleanliness of nests, egg storage, and the incubator. Additionally, the diet must be optimally suited for breeding production. An incorrect nutrient ratio or deficiencies in vitamins or minerals can have an aggravating effect on embryonic mortality. The correct care of breeding chickens, ducks, geese, turkeys, and guinea fowl guarantees not only a high hatching rate but also the viability of the young chicks. Deaths occurring in the first 14 days can often be traced back to an embryonic disease.

Diseases Resulting from Poor Hatching-Egg Quality

Not only does an infection worsen the hatching result: inadequate diet and keeping conditions also have a negative effect. If the nutrition of the embryo in the hatching egg does not function perfectly, then symptoms of illness, called "dystrophies," occur. Vitamin deficiencies principally cause such illnesses. (The particular symptoms of these illnesses are listed in the chapter on diet.) Frequently the embryos die about the eleventh day. Faulty hatching technique aggravates the course of illness even more.

Incomplete Yolk-sac Reabsorption

If the chick in the egg has drawn in the yolk sac, then it frees itself from the eggshell and closes its navel behind the yolk. The yolk is a source of nutrients that allows the chick to get by for two days without food. Its reabsorption, however, is not complete until six weeks after hatching. Until then the yolk plays an important role in fat and mineral metabolism. Disruptions in reabsorption will have a detrimental effect on development and growth. The cause of this illness can be a vitamin deficiency (especially of the vitamin-B complex) or a diet too rich in protein — that is, a to one-sided and low-quality diet. But faulty incubation (such as excessive humidity and temperature) can also lead to this abnormality. If the navel heals poorly, then too high or too low a temperature prevailed during the last days of incubation. But bacterial infection can also cause incomplete navel closing. In this case the bacteria enter the chicks and cause their illness and death.

As a preventive measure the breeding flocks must be given a diet suited to the breed. For this the breeder must know the kinds of feed and the feeding methods. In addition, artificial incubation must be done systematically, which includes the disinfection of the incubator.

Faulty Incubation

If artificial incubation is not completely controlled, then the hatching eggs will die again and again. Incubation faults can occur with respect to temperature, humidity, turning, and air supply. Depending on the intensity and duration of the incubation irregularities, slight to fatal changes in the embryo occur.

The effect of hatching defects on embryonic development are discussed in the chapter on incubation and rearing.

Illnesses Caused by Keeping Conditions

Hypothermia

Since the regulation of body temperature is not yet fully functional in chicks, they are easily subject to chilling. A long-lasting hypothermia leads to illness and finally to death. Unusual sleepiness, trembling, convulsions, and exhaustion characterize this condition.

To prevent hypothermia, the temperature should always be appropriate for the age.

Pneumoaerocystitis

By this is meant an inflammation of the lungs and air sacs of ducklings. This illness appears in particular with chilling and moistness. Weakness, lack of appetite, rattling breathing, and diarrhea are typical symptoms.

Prophylaxis: Draft, moistness, and chilling should be prevented absolutely.

Chronic Cold

Draft, moist warmth, and overly dry coop air, vitamin-A deficiency, and transportation injuries cause chronic cold.

Mucilaginous nasal secretions and the formation of a yellow, cheesy deposit in the buccal cavity and throat are typical symptoms. In severe infections the eyes bulge out. The birds pant loudly for air and try to remove the mucous. Breathing through the bill leads to drying of the tongue and to a brownish discoloration of the tip of the tongue.

The transmission of chronic cold takes place from bird to bird or through the drinking water.

Afflicted birds should be separated and given vitamins (particularly A, D, and E) as well as specific medicines. The coops must be disinfected.

Feather Picking

Feather picking should be listed among the illnesses, even though it is often designated as a "vice." As a rule, feather picking can be traced back to faulty keeping. But an incorrect diet can also cause this bad habit. There are also breeds which by nature have a tendency to display this behavior. Too limited accommodations, too dry air, strong sunlight, or oxygen deficiency promote feather picking. Usually boredom, however, is the factor responsible for this abnormal behavior. Boredom is usually brought about by a feed that satisfies too quickly or by a too unstructured coop or run. As a remedy, one offers the birds sufficient scratching space, removes surplus birds, structures the enclosure, and adds oats or oat husks to the mash feed to delay satiating the chickens. The use of styrofoam sheets or balls also provides relief. An anti-feather-pick spray available on the market has also proved effective.

Toe Picking

As with feather picking, toe picking has its cause in confined housing and a lack of diversion. If we do not use wire-mesh floors we should increase the layer of litter so that the toes always sink in the litter and are no longer visible to the other chickens. Avoiding overcrowding, a too warm coop, poor ventilation, and a quickly satiating feed also help to prevent toe picking. As a last resort we can paint the toes lightly with coal-tar. The acrid smell then frightens off the other chickens.

Nutritional Deficiency Diseases

Avitaminosis (Vitamin-deficiency Diseases)

These primarily occur among chicks and young poultry. The cause, symptoms, and remedy are covered in detail in the chapter on diet.

Rachitis

Rachitis (also called "softening of the bones") is a very damaging rearing disease. It is caused by a faulty calcium-phosphorous balance.

In rachitis the developing skeleton does not take up calcium perfectly, and insufficient bone hardness results. This leads to deformities of the legs, the wings, and the breast bone in particular.

The first indications of the disease are rough plumage, exhaustion, a slight turning in of the toes, squatting on the hock joint, and disturbances in movement. To prevent rachitis, one provides for a correct calcium-phosphorous ratio (approximately 1.5:1) by means of a mineral mixture; one also provides vitamin D. If breast-bone deformities, for example, still appear, the cause is usually too angular or too thin perches.

Perosis

Perosis is a manganese-deficiency disease; however, deficiencies in choline, niacin, zinc, and vitamin E also play a role.

First indication: With young chickens and turkeys, disturbances in growth and feathering appear. As a later typical symptom, the hock joint thickens, and the legs turn to the outside. Usually only one leg is affected.

For prevention we feed manganese in the form of manganese sulfate (0.5 grams of manganese sulfate per 10 kilograms of feed) or a mineral mixture with sufficient amounts of manganese. In addition, we should always provide poultry vitamins; a vitamin B_2 deficiency produces similar symptoms.

Toe Deformities

Toe deformities stem from hatching errors, insufficient calcium, or a diet too rich in protein. They can, however, also be the first symptoms of rachitis.

Diseases Caused by Bacteria

Pullorum

Pullorum attacks chicks of the age of up to two weeks. Pullorum is caused by the bacterium *Salmonella pullorum.*

If the pathogen is already present in the egg, then the chicks usually die before hatching. If infected chicks still manage to hatch, then they stand huddled in the brooder — as if they are freezing — with drooping wings and closed eyes. The typical symptom is a chalk-white diarrhea, which leads to the encrustation of the down around the vent with droppings of the same color. Healthy chicks become infected very quickly and perish somewhat later than the others. An additional symptom is disturbed breathing.

Surviving chicks grow up to be adults, but they carry the bacterium hidden inside themselves and often transmit the pathogen through the hatching eggs.

To combat the disease, one culls infected birds if necessary, or administers specific medications to them as well as the ones that appear to be healthy.

Poultry Cholera

Bacterium avisepticum, which is unrelated to the human cholera pathogen, causes poultry cholera. Waterfowl are much more likely to become infected with this disease than chickens.

In poultry cholera the head points discolor to a blue-red, and the diarrhea which sets in has a greenish-yellow color, is interspersed with flecks, and has an unpleasant odor. Difficulty in breathing and convulsions often appear. Often the birds die overnight without previous symptoms of any kind. Infection takes place through the droppings or through infected drinking water and feed.

Disinfection of the coops and a thorough veterinary treatment with sulfonamides and antibiotics can still save the infected birds.

Tuberculosis

This disease — like poultry cholera — is caused by bacteria. The pathogens are discharged in the droppings and are transmitted through infected feed or water, or through direct contact with droppings. Poultry tuberculosis is also infectious to other species of domestic birds and humans. Emaciation, dull plumage, blue-red or gray-red comb discoloration, and symptoms of diarrhea appear. Usually the face also becomes wasted.

After an infection we thoroughly disinfect the coop with appropriate agents. Infected birds are slaughtered immediately.

Paratyphus

Paratyphus is a *Salmonella* disease. Primarily chicks are infected with this disease. The symptoms externally resemble those of pullorum. In rare cases blindness in one eye occurs.

With waterfowl it appears as unkempt plumage, loss of vitality, and thirst. Later, disturbances of balance and convulsions occur. The end is usually not long in coming.

Specific medications such as antibiotics and sulfonamides can bring about a cure if treated in time.

Protozoal Diseases

Coccidiosis

This disease typifies transmission through contact between birds and droppings. Coccidia are protozoan parasites which live in the intestinal tract of chickens, turkeys, and geese. Their eggs develop in the outside world. There they can survive for up to a year, especially in a wet environment. Chicks up to the eighth week of life, in particular, are infected. In this illness their damaged caecal mucous membrane discharges blood with the usually watery droppings. After the eighth week the susceptibility to caecal coccidiosis is reduced, but the susceptibility to small-intestine coccidiosis remains and can also appear well into the pullet stage. General symptoms of illness are weaknesses, lethargy, and lack of appetite. Coccidiostats are contained in the chick and grower complete mash found in the trade. Whoever does not use this feed should, as a preventive measure, provide the birds with coccidiostats in the drinking water or through forced feeding (particularly for therapy). The prepared feed only prevents a moderate coccal invasion, however. For a massive infestation only veterinary medicines are called for. One immediately places infected birds in a separate coop that has a full wire-mesh floor, and treats the infected and uninfected birds with a specific remedy (for example, a sulfathiazole preparation).

Blackhead Disease

Blackhead disease is an illness that primarily appears in turkeys. The infected birds have unkempt plumage and stand around listlessly. The diarrhea which occurs is yellowish to sulfur-yellow and has a liquid consistency. The head sometimes exhibits a bluish to black discoloration.

Special preparations are used for cure and prevention.

Viral Diseases

Diphtheria and Fowl Pox

Fowl pox and diphtheria are produced by the same virus. If the skin is infected, pox results; in the case of mucous-membrane infection, diphtheria results. With fowl pox the unfeathered areas of skin such as the wattles, ear-lobes, comb, and corner of the beak swell up. The affected areas of skin at first show a yellowish, and later a dirty-brown, scab-covered pigmentation. The pox disease, as a rule, is benign, and the pox disappear on their own.

Diphtheria occurs principally in youngsters. The beak, larynx, and windpipe exhibit yellowish-white deposits. In contrast to chronic cold, the deposits grow into the mucous membrane. The birds frequently gasp for air with open beaks.

Special preparations and an inoculation help to fight or combat diphtheria.

Marek's Disease

Marek's disease is an infectious viral disease of youngsters. The virus attacks the nervous system and brain. The afflicted birds exhibit poor leg coordination and die with one leg stretched to the front and the other to the rear. As a result of the nervous paralysis, the pupils exhibit an irregular border and there is a gray-green iris pigmentation. The wings usually hang down from the body.

One achieves effective control through the vaccination of one-day-old chicks.

Classic Fowl Pest

As a rule, this appears very suddenly. Symptoms of this disease are uncoordinated gait, exhaustion, unkempt plumage, and weakness. Furthermore, greenish diarrhea and difficulty in breathing occur. A preventive inoculation by a veterinarian is possible. The disease is no longer acute at the present time.

Newcastle Disease

Infected chicks show difficulty in breathing and soon perish. Adult birds also suffer from difficulty in breathing and often cough out a mucilaginous secretion which only sticks slightly to the beak. Lack of appetite and weakness are clearly recognizable. It often produces lameness and convulsions and a twisting of the head of up to 180 degrees (the last-named symptom can also be observed in vitamin-deficiency illnesses). Preventive control results by means of an inoculation through the drinking water. At most German poultry shows, proof of vaccination is required.

Leucosis

Leucosis is not an exactly definable disease, since the virus attacks various cellular tissues. Although the bird often appears to be externally healthy, it can conceal and excrete the virus, particularly through the egg and the droppings. Primarily chickens and turkeys are infected. The symptoms of this disease are not distinctive. Frequently the comb is pale and shriveled. Diarrhea, lack of appetite, and emaciation are additional symptoms. Interestingly, high-production breeds with intensive protein feeding are most likely to suffer from leucosis.

So far, only preventive keeping and care can be recommended; infected birds should be destroyed. With time the immune system of the poultry develops antibodies which are able to neutralize the viruses. For this reason, with birds susceptible to leucosis only those individuals two or more years of age should be used for breeding. In the incubator one should always hatch only chicks of a single breed, since the chicks of different breeds can infect one another if a bird carries the leucosis virus inside itself. After incubation the incubator should be appropriately disinfected each time.

Worm Infestations

Threadworm Infestation

The 1–2-centimeter worms inhabit the small intestine, particularly in young poultry, and bore into the intestinal mucous membrane and suck blood, producing anemia.

Heavily infested birds exhibit paleness, signs of emaciation, and diarrhea; the diarrhea often causes the feathers around the cloaca to become soiled with droppings. Occasionally, wing and leg weakness also occurs.

Specific remedies help to control this illness. Usually a preventive onion-and-garlic diet prevents the massive spread of threadworms.

Tapeworm Infestation

The tape worm is a roundworm approximately seven centimeters long. This parasite lives in the intestinal mucous membrane and bores through it at times, causing great damage.

Here too we use special worm-control remedies if necessary; generally, we preventively feed onion and garlic.

With heavily infested birds, which exhibit the same symptoms as birds infested with threadworms, one can occasionally find dead tapeworms in the droppings.

Diseases Produced by Ectoparasites

Skin and feather parasites are injurious to the vitality of our poultry. Birds that are not perfectly healthy seldom produce satisfactory offspring, and often even fertility leaves something to be desired. The culprits are usually feather mites, mites, ticks, and fleas.

Feather-Mite Infestation

Feather mites, which feed on skin scurf, parts of feathers, and glandular secretions, are undoubtedly the primary parasite. Their presence is revealed by holes in the feathers. A rough and disarranged plumage results. In addition, the occurrence of feather mites is revealed by egg masses on the feather shaft and the base of the feather. The principal sites where the eggs are laid are the head, the belly, the area under the wings, and especially the feathers around the vent. With chickens, infestation is apparent from restless behavior and constant pecking and scratching. A heavy infestation in adult birds causes emaciation or a reduction in laying production. With youngsters, disturbances or stoppages in growth occur. Particularly during the molt, one must make sure that the chickens are free of feather mites, since otherwise the feathers will take too long to grow back. Poultry can be protected against infestation by dusting with a powdered insecticide safe for use on the skin.

Mite Infestation

Another kind of parasite is the blood-sucking bird mite. In contrast to feather mites, which spend the entire day on the bird's body, bird mites spend the day in cracks and dark corners of the coops and only infest the birds at night, during which time they feed on their blood. With a heavy infestation, the chicken usually stops laying, or even death can occur. Infested birds can, as a rule, be recognized by their pale combs. In addition, emaciation, feather loss, and anemia appear. If the parasite establishes itself in the auditory canal, disturbances in balance occur.

If wing and tail feathers lose their luster and fall out after a time, this is an indication of infestation by feather-quill mites.

The familiar "scaly legs" of chickens are caused by mange mites.

The mites establish themselves under the scales of the legs; the consequence is growths of the skin tissue, which finally lead to the formation of a thick, scabby layer — the scaly legs. For prevention, the coops are limed; afflicted birds must be treated according to directions with scaly-leg control remedies (ointments) which are available on the market. As a general control measure against mites, a thorough coop disinfection should be carried out first of all. The perches especially must be painted with creosote and rinsed with a disinfectant.

Bird-Tick Infestation

Bird ticks, which belong to the same order as mites, also suck blood. The ticks seek out the bird during the night, while they remain in their hiding places during the day. Their sucking sites on the poultry are recognizable through a red puncture with a blue-red border. Preferred regions are the nape, thighs, and the inside of the wings. Otherwise the same symptoms as with a mite infestation occur. Poultry plagued by bird ticks can also die as a result of the great loss of blood, particularly youngsters. Affected birds are rubbed with oil at the appropriate places; the mites are subsequently removed with forceps. If the breeder does not oil the affected site and the tick before trying to remove the tick, its head will break off and produce a purulent inflammation on the poultry.

Flea Infestation

Fleas lay their eggs in the litter and droppings. After about ten days, the larvae hatch from the eggs and subsequently pupate in the coop litter or in the nest. After about twenty more days the fleas hatch and infest the poultry. As a consequence, the poultry seek out other nests, which can often lead to mislaying of eggs. A countermeasure consists of dusting the birds and nests with a safe powdered insecticide.

Natural Control of Ectoparasites

As a general principle, the breeder should provide his birds with a dust bath (in the case of chickens) or a bathing tub or pond (in the case of waterfowl). These will enable the poultry to keep themselves free of vermin in a natural manner. A supplement of powdered insecticide in the dust or sand bath strongly supplements its effect.

In general, considerable importance must be attached to the control of parasites, since otherwise damage that cannot be made good can occur.

A FURTHER STEP: ORGANIZED BREEDING

Membership in a Poultry Association

If one has found an interest in poultry keeping or breeding, then he can join a regional poultry association, the goal of which is to breed poultry according to very specific, established breed criteria. Besides beauty and the bird's reproductive ability, the poultry breeder — depending on the breed — of course also attaches importance to production. In the association an animated exchange of ideas with respect to the problems of breeding and keeping the most diverse breeds prevails. If one is no longer making progress in breeding, then a fellow member often has advice. In the association comradeship reigns, and many friendships have developed in a poultry-breeding association. In the exhibition season, the association as a rule organizes a show, which displays the association and its breeding accomplishments to the outside world. As a result, publicity for poultry breeding is provided, and many city children experience close contact with the living bird. Beyond that, poultry breeding ensures the preservation of the often antique cultivated breeds in a suitable setting. Domestic animal breeding is a piece of living cultural history and therefore merits just as high an opinion as many other culturally significant achievements of mankind.

So that this task can be optimally fulfilled, in Germany some associations make a tract of land within the limits of a breeding facility available to the breeders. In return, the association expects the active exhibition of high-quality poultry breeds. Particularly in the city, breeding facilities of this kind are veritable oases in the concrete desert. Unfortunately, only a handful of poultry associations have their own breeding facilities, so that many breeders must find their own suitable tract of land.

In addition, the association also offers good company. Monthly meetings, summer festivities, excursions, or gatherings of breeders offer sufficient opportunity to forget the bustle of everyday life for a while.

It should not go unmentioned that joining the association entails the responsibility of duties. Annual dues must be paid, and sometimes one must place one's services at the association's disposal at seemingly inconvenient times, particularly if one has accepted an executive office. Nevertheless, even an obligation of this kind provides great pleasure if one can harvest the fruit of what the group accomplishes.

The association is directed by various people. At the top stand the president and vice-president, as well as the secretary and the treasurer. Various other officials are also present on the board of directors, depending on need. The press secretary, who also represents the association externally, undoubtedly has the most important function in this connection.

Competition at Shows

At shows the breeder presents the work he accomplished in the previous year. Depending on the quality of the parent birds, the management of the breeding line, and the breeder's skill, the birds receive higher or lower marks. A judge evaluates the exhibited poultry breeds and sometimes writes a critique of the breeding strain. The beginner should not be discouraged by possible initial lack of success. With a little experience and the help of breeder friends, one soon arrives at the qualified selection of exhibition birds — and thus at success. Here too, no man is born a master of his craft, and one certainly cannot expect to contribute top show birds in the very first year, but continual ascent can be predicted with concentrated and competent breeding effort with first-rate birds.

If you have earned a Third, or even a First Place, then you can be completely satisfied. Some First Place birds even receive a prize. However, it should not go unmentioned that you must pay an entry fee for each bird at shows, so that you break even only rarely at large shows. Thus you cannot make a "business" out of it, but you and your stock will become known throughout the country, and you will constantly enlarge your circle of acquaintances and friends. The shows thus have more intangible than material value. But whoever is gripped by exhibition fever will not be able to get rid of it so easily.

Poultry for Competition

In order that your birds stand out well at shows, they must fulfill very definite criteria. What these amount to in practice will be briefly described here.

Poultry which are exhibited at shows must be in full plumage, as pertains to the specific breed. In addition, all characteristics of the breed must be present. These are listed for each poultry breed in the Poultry Standard of the various national poultry associations. The judges make their evaluations according to it. All technical terms are listed in the standard. Additionally, the poultry must not exhibit deformed toes or other malformations of the bones. The shape of the comb must agree exactly with the standard. Faults of the plumage are penalized. If a bird is quite pale — whether because of disease or lack of condition (insufficient age) — one should leave it at home. Likewise, poultry with parasites are not presentable.

Before the show, one cleans the legs and toes in warm, soapy water with a hand brush. After drying, rub them with skin oil, cream, or petroleum jelly. As a result, the bird looks cared for and clean at the show. One washes dirty combs and facial areas with clean, cold water. Subsequently the head areas are oiled like the legs. The plumage should not be washed before the show since the feather structure is often damaged in this way. With regular care, washing the plumage is unnecessary. White and light color varieties, however, are an exception. With them, one must often turn to the wash tub despite the best practices. It is important that the washing take place in a heated room. Washing is done with water containing soap or a detergent. Washing is done in the direction of the feathers. The final, thorough rinsing with clear, slightly acidic water should be given careful attention. Drying, which should also take place in the direction of the feathers, hastens drying off. After washing we keep the birds on clean, dry litter so that they do not immediately become soiled again. For shipping the birds to the show, one chooses a transport container that is the right size for the poultry in question. The birds must still be able to move in the shipping crate, but only enough that they will not damage any feathers, since this would reduce their exhibition value.

Care of Poultry After the Show

When the birds come back from the show, they require special care. From time to time one of them will catch a cold. An infected bird must be kept in quarantine and nursed to health with medications. A dose of vitamins given to all of the birds immediately after returning helps to counter the stressful situation of the show. With roosters especially, a return to the flock will no longer be possible, since the birds have become strangers to one another, no longer recognize their old positions in the pecking order, and fight for a new one. In the process many birds lose comb points and have pecked, bloody faces as well as frayed wattles. Such roosters can of course no longer be used for future shows. For this reason, roosters after the show are placed in separate cages, which offer them ample freedom of movement. Depending on the breed, this method of keeping is also recommended for hens. Particularly with bearded birds, fighting upon reintroduction may result in loss of the beard feathers; for such chickens separate housing between shows should be strived for. Moreover, we should not exhibit the same bird all too often, since its condition will likely deteriorate considerably. A regular examination of the exhibited birds for parasites is advisable, since the poultry bring back undesirable parasites from shows from time to time.

Keeping Breeding Records

So that we do not leave breeding to chance, but with the aid of the fundamentals of genetics instead work systematically, we must keep records. At first hearing this sounds more difficult that it really is. A few notations are all one needs to breed systematically.

As the starting point, we take the parent birds and write brief notes about each, stating the strengths and weaknesses. As a distinguishing mark for each bird, it is most practical to use the band number. By means of trap-nest inspection and the marking of chicks, we know the identity of all the chicks from each hen. If we examine their external appearance, then we can recognize quite soon how the parent birds have passed on their genes. If the inheritance picture is positive, then we can continue to work in this line with the best birds. If we make such records every year, we obtain an exact record of kinship with great-grandparents, grandparents, parents, and the most recent generation. If unexpected characteristics suddenly turn up, then we can trace these back with the aid of our records. With the information from careful genealogical control, the genotype, which we cannot see in the phenotype, will become comprehensible, as long as we work in one line, that is to say, mate only closely related birds with one another (for example, son and mother, or brother and sister).

The records do not require much work, yet they simplify breeding to a considerable degree and make it possible for us to quickly work to the top of purebred-poultry breeding. In so doing we will quickly surpass the army of poultry breeders who leave breeding more or less to chance.

The Sale of Eggs, Day-old Chicks, Youngsters, and Breeding Birds

Hatching eggs may only be sold if they come from purebred (this does not apply to hybrids), satisfactorily grown, and healthy breeding birds. Appropriate offers are to be clearly written, and the delivery must take place in accordance with the offer. Hatching eggs may not be older than eight days at the time of shipment. They must have been properly stored until they were shipped and must carry the identification mark of the seller. Additionally, they must be undamaged and clean (unwashed) and have normal shell formation. The minimum weight must be adhered to. The cost of shockproof packaging can be added to the seller's normal price. A normal fertility rate of 75% with light and medium-weight breeds and 50% with heavy breeds applies. Complaints must be registered within 14 days.

Like hatching eggs, chicks must also come from purebred (this again does not apply to hybrids), satisfactorily grown, and healthy breeding birds. The seller accepts the responsibility that the chicks will be alive upon arrival, and has to provide for replacements for those that arrive dead, provided official confirmation of the loss from the carrier is presented. The replacement can also be in the form of money returned.

If we deliver youngsters or breeding birds, then these should meet the characteristics of the breed to a high degree and should be in vigorous and healthy condition. If we order birds, sometimes we receive poor stock; we should send these birds back. For an expert opinion in questionable cases we should ask a competent breeder from the association for help. Birds that were seriously ill or whose outward appearance was altered through medical, chemical, or physical action may never be sold, since this would be intentional fraud. When selling birds we should start with the principle that we should only sell birds we would also wish to buy ourselves. We can, of course, expect to obtain a good price for a good bird.

Whether we have bought one or several birds, we often have acclimation difficulties. Fights for position in the hierarchy occur between them and the "old, established" birds. As a rule the new birds are defeated and thereby exposed to social stress. To reduce this, we house the new birds separately, but adjacent to the established ones. In this way, the new arrivals become familiar with the strange surroundings and the established poultry, and can hold their own substantially better during introduction later, which will guarantee their active incorporation in the daily routine of the flock.

Measurement Conversion Factors

When you know—	*Multiply by—*	*To find—*
Length:		
Millimeters (mm)	0.04	inches (in)
Centimeters (cm)	0.4	inches (in)
Meters (m)	3.3	feet (ft)
Meters (m)	1.1	yards (yd)
Kilometers (km)	0.6	miles (mi)
Inches (in)	2.54	centimeters (cm)
Feet (ft)	30	centimeters (cm)
Yards (yd)	0.9	meters (m)
Miles (mi)	1.6	kilometers (km)
Area:		
Square centimeters (cm^2)	0.16	square inches (sq in)
Square meters (m^2)	1.2	square yards (sq yd)
Square kilometers (km^2)	0.4	square miles (sq mi)
Hectares (ha)	2.5	acres
Square inches (sq in)	6.5	square centimeters (cm^2)
Square feet (sq ft)	0.09	square meters (m^2)
Square yards (sq yd)	0.8	square meters (m^2)
Square miles (sq mi)	1.2	square kilometers (km^2)
Acres	0.4	hectares (ha)
Mass (Weight):		
Grams (g)	0.035	ounces (oz)
Kilograms (kg)	2.2	pounds (lb)
Ounces (oz)	28	grams (g)
Pounds (lb)	0.45	kilograms (kg)
Volume:		
Milliliters (ml)	0.03	fluid ounces (fl oz)
Liters (L)	2.1	pints (pt)
Liters (L)	1.06	quarts (qt)
Liters (L)	0.26	U.S. gallons (gal)
Liters (L)	0.22	Imperial gallons (gal)
Cubic centimeters (cc)	16.387	cubic inches (cu in)
Cubic meters (cm^3)	35	cubic feet (cu ft)
Cubic meters (cm^3)	1.3	cubic yards (cu yd)
Teaspoons (tsp)	5	millimeters (ml)
Tablespoons (tbsp)	15	millimeters (ml)
Fluid ounces (fl oz)	30	millimeters (ml)
Cups (c)	0.24	liters (L)
Pints (pt)	0.47	liters (L)
Quarts (qt)	0.95	liters (L)
U.S. gallons (gal)	3.8	liters (L)
U.S. gallons (gal)	231	cubic inches (cu in)
Imperial gallons (gal)	4.5	liters (L)
Imperial gallons (gal)	277.42	cubic inches (cu in)
Cubic inches (cu in)	0.061	cubic centimeters (cc)
Cubic feet (cu ft)	0.028	cubic meters (m^3)
Cubic yards (cu yd)	0.76	cubic meters (m^3)
Temperature:		
Celsius (°C)	multiply by 1.8, add 32	Fahrenheit (°F)
Fahrenheit (°F)	subtract 32, multiply by 0.555	Celsius (°C)

Index